PRAISE FOR *Gods, Wasps and Stranglers*

"This book concerns the stunningly versatile and ancient family of fig trees now being used as a framework species to restore damaged tropical forests. Figs are not only considered the keystone species in forests but are perhaps the world's most perfect tree—they provide highly nutritious fruits with health-giving and medicinal qualities. They attract birds and animals. They grow very rapidly and produce abundant fruits in a few years. They make shade and shelter, their deep powerful roots can break up compacted soils, they draw up water, they prevent erosion, and they have important spiritual qualities. The tree in the Garden of Eden was very likely not an apple but a fig."

—Annie Proulx

"Surprising, engrossing, disturbing, and promising, *Gods, Wasps and Stranglers* combines masterful storytelling and spellbinding science. This is a beautifully written and important book about trees that have shaped human destiny."

—Sy Montgomery,
author of *The Soul of an Octopus*

"The complex web of ecological connections between fig trees, tropical forest animals and plants, as well as people and human culture is nothing short of a marvel. *Gods, Wasps and Stranglers* is a page-turner and a revelation: You will never again think of a fig as just something to eat. There is no better way to introduce the complexity and wonder of nature—and our intricate relationship with it. A must read."

—Thomas E. Lovejoy, University Professor of
Environmental Science and Policy, George Mason
University; fellow, National Geographic Society

"In *Gods, Wasps and Stranglers*, rainforest ecologist Mike Shanahan charts a lifelong love affair with figs, one that has taken him from India to Kenya, through temples and rainforests, all in search of a deeper understanding of what he describes as 'humanity's relationship with nature.' The fig becomes a tasty lens that reveals not only the fruit's cultural and biological significance but our relationship to that which most deeply nourishes us."

—**Simran Sethi**,
author of *Bread, Wine, Chocolate*

"A real labour of love, concisely and elegantly told."
—**Fred Pearce**, author;
environmental consultant, *New Scientist*

"In his insightful book, *Gods, Wasps and Stranglers*, Mike Shanahan combines poetry and science, history and humanity, to tell a story not only of the fig tree but of life on Earth in all its beautiful and astonishing complexity. In doing so, he reminds us of what a remarkable place we inhabit—and how much we should all want to protect and preserve it."
—**Deborah Blum**, director of
Knight Science Journalism Program, MIT;
author of *The Poisoner's Handbook*

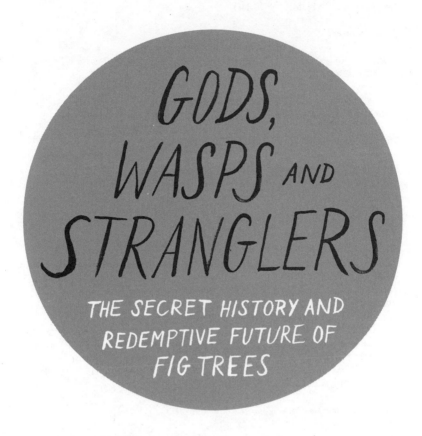

GODS, WASPS AND STRANGLERS

THE SECRET HISTORY AND REDEMPTIVE FUTURE OF FIG TREES

MIKE SHANAHAN

Chelsea Green Publishing
White River Junction, Vermont

Originally published in the United Kingdom
by Unbound in 2016 as *Ladders to Heaven*.

This edition published by Chelsea Green Publishing, 2016.

Illustrations by Mike Shanahan.

Printed in the United States of America.
First printing October, 2016.
10 9 8 7 6 5 4 3 2 1   16 17 18 19 20

Our Commitment to Green Publishing
Chelsea Green sees publishing as a tool for cultural change and ecological stewardship.
We strive to align our book manufacturing practices with our editorial mission and to
reduce the impact of our business enterprise in the environment. We print our books and
catalogs on chlorine-free recycled paper, using vegetable-based inks whenever possible.
This book may cost slightly more because it was printed on paper that contains recycled
fiber, and we hope you'll agree that it's worth it. Chelsea Green is a member of the Green
Press Initiative (www.greenpressinitiative.org), a nonprofit coalition of publishers, man-
ufacturers, and authors working to protect the world's endangered forests and conserve
natural resources. *Gods, Wasps and Stranglers* was printed on paper supplied by
Thomson-Shore that contains 100% postconsumer recycled fiber.

Library of Congress Cataloging-in-Publication Data
Names: Shanahan, Mike, author.
Title: Gods, wasps, and stranglers : the secret history and redemptive future of fig trees
  / Mike Shanahan.
Description: White River Junction, Vermont : Chelsea Green Publishing, 2016.
Identifiers: LCCN 2016039240| ISBN 9781603587143 (hardcover) | ISBN
  9781603587150 (ebook)
Subjects: LCSH: Ficus (Plants) | Human-plant relationships.
Classification: LCC QK495.M73 S53 2016 | DDC 583/.45--dc23
LC record available at https://lccn.loc.gov/2016039240
Library of Congress Cataloging-in-Publication Data

Chelsea Green Publishing
85 North Main Street, Suite 120
White River Junction, VT 05001
(802) 295-6300
www.chelseagreen.com

MIX
Paper from
responsible sources
FSC
www.fsc.org   FSC® C013483

For Charlotte and Noah

Then the trees said to the fig tree
'You come and be our king!'
The fig tree replied,
'Must I forgo my sweetness,
forgo my excellent fruit,
to go and sway over the trees?'
    Judges 9: 12-13 (Old Testament)

'I have not cut down any fig tree . . . why then does
calamity befall me?'
    Ravana, the ten-headed demon-king of Lanka, in
    The Ramayana (c. 500-400 BCE)

'Every fruit has its secret. The fig is a very secretive
fruit.'
    DH Lawrence

Contents

List of Illustrations

ONE:

Snakes and Ladders
& Tantalising Figs

The figs were big orange beacons that lured me from afar. The snake was lime green and venomous and just centimetres from my face. I found them both near the top of a tall tree in a Bornean rainforest. While the snake was safely coiled on a sturdy branch, all I had were some sweat-soaked fingers to save me from a fall. My heart raced. The snake's unblinking eyes looked as patient as time.

The year was 1998, and I was falling headlong into a fascinating story. The stars of the story are the fig trees — the 750 or so *Ficus* species. Over millions of years these trees have shaped our world, influenced our evolution, nourished our bodies and fed our imaginations. The best could be yet to come. Fig trees could help us restore ravaged rainforests, stem the loss of wild species and even limit climate change. They could build vital bridges between scientific and faith-based world views. Their story reminds us of what we all share. It warns us of what we could lose.

In Greek mythology, a branch laden with sweet figs was among the temptations that teased the demigod Tantalus during his punishment in the Underworld. Each time Tantalus reached for the figs, the wind wrenched the tree's bough beyond his reach. This tale gave English the verb 'to tantalise'. Those dull orange figs in Borneo, with their guardian snake, seemed certain to elude me too. I craved them, though I had no desire to eat their flesh.

The figs adorned the stubby branches of a *Ficus auranti-aca*, a species that starts out in life on the rainforest floor and climbs up the trunk of a big tree, growing as straight as a charmed snake. As it rises, it paints its host tree's bark green with thousands of little leaves. This species relies heavily on primates to eat its figs and disperse the tiny seeds within them. But in this particular forest, *Ficus aurantiaca* was a plant with problems.

First, few primates remained in the area. The national park I was in was an island of ancient forest, at whose edges lapped a biting tide of oil palm plantations, farms and logging concessions. These man-made habitats posed dangers to primates and other wildlife. Like the national park itself, they were often visited by shotgun-toting hunters for whom a monkey or a gibbon would be a prize kill. But even if primates had been abundant, something else was wrong. Like all fig species, *Ficus aurantiaca* depends on tiny wasps to pollinate its flowers. That year, however, an intense drought had stricken the national park. By forcing *Ficus aurantiaca* to stop producing figs, the drought had driven these insects into local extinction.

Across the national park many other *Ficus* species faced the same problem. Without figs and fig-wasps, the plants could not reproduce. The many bird and mammal species that rely on figs as food faced a serious fruit shortage. And that would have knock-on effects for the other plants whose seeds these creatures dispersed. That's why I was so interested in the *Ficus aurantiaca* figs I spotted that day. They were a sign that life might be returning to normal. But to be sure, I needed to know what was going on inside them. All that lay between my curiosity and the answers I sought was that venomous snake and the risk of a long fall.

I was hanging from the last of seven ladders that some-body had lashed, toes-to-shoulders, flush to the tree. I had not brought a safety harness but climbed nonetheless. Those figs had banished security from my mind. As I tried to maintain my grip on the metal rungs, an intense wave of vertigo paralysed me. For a moment I could no longer sense my own body. It was as if my mind was all that existed of me, and it scrambled to process the sudden danger. I held my breath. Solid ground was a long way down, but I needed those figs and that meant I had to let go of the ladder with one hand to reach past the snake.

My obsession with figs had long since taken root. In the years ahead it would take me to temples, mountaintops and into the crater of an active volcano. It would stay with me long after I stopped studying biology. It would take solid form in the pages of this book. It has been a long and fruitful journey.

I first encountered these plants in my childhood home. One lived inside in a plastic pot and inched its way towards whatever light it could find. Just a metre high and spindly of limb, it would fall over if my sister or I ran past it too quickly. I was the youngest member of the family, so the periodic task of cleaning the tree's leaves fell to me. As I stroked each dark green leaf with a soft yellow duster so it shone, I couldn't fail to notice the label that stood upright like a tombstone in the soil. The exotic words there—*Ficus benjamina*—made no sense to my young eyes. Years would pass before I understood what those words meant and before I learned that this frail plant was a mere baby.

As one of the most common houseplants in the world, *Ficus benjamina* is seen more often indoors than out. But in the forests of Asia, this species can reach 30 metres in height. Its reddish pea-sized figs sustain dozens of wild animal species. My mum referred to the plant in our hallway as 'the fig tree'. This confused me, as outside, in our next-door neighbours' garden, there was another plant she gave the same name though it looked utterly dissimilar. Despite their differences, both plants were indeed fig trees, distant cousins with a common ancestor. This was my first exposure to the rich variety of fig species that has kept biologists busy for more than 2,300 years.

The leaves of the plant inside were small and smooth, but those of the tree outside were rough to touch and wider than my dad's handspan. Unlike the tiny tree inside, the one outside was big enough to peer down at me from behind the two-metre tall wall that separated our garden from our neighbours'. And unlike the barren houseplant, the one outside often tempted me with its fruit—soft and tasty figs. It was *Ficus carica*, the so-called edible fig. The ancient Greeks valued this species so much they believed it to be a gift from the gods.

These fig trees sowed in my young mind some seeds that would long lay dormant. In time they germinated into a garden of fascination. Fig trees would show me the world through different eyes and different taste buds—those of diverse cultures from the present day and the distant past, as well as those of bats and birds, monkeys and much stranger beasts. This germination began in 1994, at the University of Leeds, in a lecture by biologist Steve Compton. He would later list his research interests on the university website as including 'anything to do with fig trees and fig-wasps'. I had never heard of a fig-wasp. Steve changed my life when he taught me their story. It's a story to which DH Lawrence alluded in his poem about the 'Wicked fig tree' with its 'self-conscious secret fruit'.

Lawrence hinted at the trait that characterises all *Ficus* species and explains why they matter so much. It relates to sex. For any species of flowering plant to reproduce sexually, male pollen must fuse with female ovules—just like the sperm and eggs of mammals. The fertilised ovules turn into seeds. These are plant embryos. They develop inside their mother flowers until they are ready to disperse and take their chances in the game of life.

Some plants rely on the wind to transfer their pollen, but the vast majority need help from animals. Many rely on insects such as beetles, bees and butterflies, which spread pollen as they wander from flower to flower. More rarely, birds or mammals provide the service. Bats that feed on nectar pollinate the flowers of the blue agave, the succulent plant that is the source of tequila. Their pollen-dusted noses fit into the agave flowers like hands into gloves. Figs and their fig-wasp pollinators have an even tighter relationship, but have you ever seen flowers on a fig tree? I think not.

Sit under a *Ficus* tree every day of your life and you will see it produce crop after crop of figs but never a single flower. The mystery of this apparent virgin birth has endured for millennia. It is why ancient Hindu and Buddhist texts use the phrase 'seeking flowers in a fig tree' to describe a hopeless search. It's why a Bengali saying describes someone who has become 'invisible like a fig flower'. And why the Chinese characters for fig—無花果—mean 'the flowerless fruit'. The ancient phrase-makers never looked closely enough.

A proverb in the Tamil language of southern India comes close to solving the mystery. It refers to the way a fig tree's 'flowers bloom secretly and fruits flourish visually'. Here is the secret: the fig is not a fruit at all. It is a hollow ball whose entire inner surface is lined with tiny flowers that never see daylight.

The mystery of the hidden flowers arose as many as 80 million years ago when fig trees and fig-wasps forged a fortuitous bond that became biological shackles for them both. Fig flowers were once out in the open but, as the ancestors of today's *Ficus* and fig-wasp species developed their codependence, the plants evolved to exclude many other species from their flowers. Over generations, the platforms upon which the flowers stood developed into urn-like figs that hid the blooms away. Since then figs and fig-wasps have spread across the planet and diversified into an astonishing variety of species.

Today, every one of the 750 plus *Ficus* species is pollinated by just one or more tiny wasp species that can feed and breed only in their fig partner's flowers. The *Ficus* and fig-wasp species cannot survive without each other. Their destinies are bound together, and many more species gain as a result. I know because I counted just one group of beneficiaries and it took me years. To find out which animals eat figs, I trawled through hundreds of scientific papers, recorded my own observations and begged biologists to share unpublished data. The list that emerged was more motley than I could have imagined.

It included elephants and opossums, bearded pigs and spectacled bears, mongooses, bandicoots and tree kangaroos. I found records of 90 species of fruit bat and 80 species of primate eating figs. Deer, cattle and antelopes eat them. So do rats, mice and more than 30 species of squirrel. I even found records of fig-eating by giant tortoises and jackals, kingfishers and seagulls and several species of fish. The list included hundreds of species of birds, from pheasants to orioles, ostriches to woodpeckers and more than 120 species each of parrots and pigeons. In all, I found that figs feed at least 1,274 species of birds and mammals—

7

far more than any other type of fruit. These animals in turn disperse the seeds of thousands of other plant species. So figs help to sustain life across entire landscapes, and it is because of this that I came to be in Borneo, hanging from a ladder high up a tree.

Tantalus never got his figs, but I got mine. As I plucked each one and shoved it into my pocket, sticky white latex oozed from the wounds I inflicted. This gummy sap is common to all fig species. It deters insects from feeding on them, and it speeds recovery from injuries. For millennia, people have found ways to make fig latex work for them too. They have smeared it on branches to catch birds, curdled milk with it to make cheese, processed it to make rubber and used it as an aphrodisiac. I was glad of the latex that day. It covered my hands and strengthened my grip as I lowered myself down the chain of ladders and entered the murky forest again. The snake had not moved a single scale.

Back at my workbench in my research station in the national park, I picked up a sharp scalpel and sliced into each fig. One by one they opened like mouths with nothing to say—they were barren hollows. Their flowers were unpollinated. They bore no seeds. Their pollinator wasps had not returned. My climb had been in vain.

It may seem strange to take risks in search of rare figs, but by the time I met that snake in 1998, I had come to understand three things: that fig trees play important roles in rainforests; that they have influenced diverse human cultures; and that today we are both destroying the former functions and forgetting the latter. Back then the science was fascinating enough. But I was learning too about a chain of reverence towards fig trees that coils back in time for many thousands of years and encircles the globe.

Fig trees appear in mythologies in the Amazon and in Africa, across the Mediterranean and the Middle East, and from the foothills of the Himalayas to the islands of the South Pacific. They feature in some way in every major religion, starring in the stories of Krishna and Buddha, Jesus and Muhammad. But these are all just recent examples. Fig trees were inspiring, sustaining and even protecting our ancestors long before they invented writing or domesticated the dog. They were among the first plants people cultivated, and their figs are among the most nutritious of foods we eat today. Thanks to their curious biology, arguably no other group of plants holds as much ecological importance, evolutionary interest and cultural value.

Our shared story stretches back beyond the birth of our species, back to before our ancestors descended from the trees and walked upright for the first time. It is a story of life and death and of a deal undone. Its cast includes kings and queens, gods and prophets, flying foxes and 'botanical monkeys'. It features scientific and religious wonders born from biology that seems almost impossible in its elegance. Most of all it is a story about our relationship with nature. The story stretches back tens of millions of years to the age of the giant dinosaurs, but is as relevant to our future as to our past. As our planet's climate changes and reminds us that nature really does matter, the story has important lessons for us all.

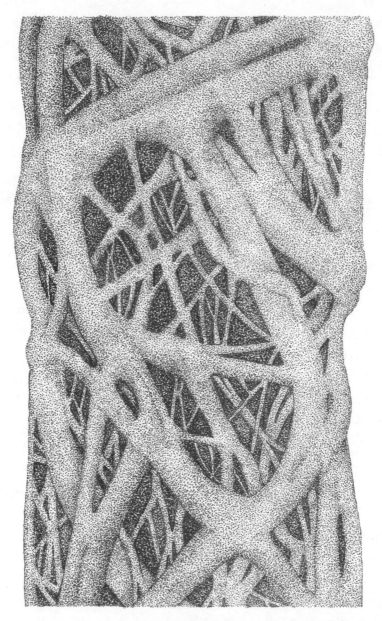

DEADLY EMBRACE:
A strangler fig imprisons its host tree with its rampant aerial roots

TWO

Trees of Life, Trees of Knowledge

On a July morning in 1860, Alfred Russel Wallace woke beneath a ceiling of woven palm fronds in what was little more than a shed, his temporary home in a rainforest just south of the equator. Wallace was on Waigeo, an island in what is now eastern Indonesia. As the forest creatures called forth the new day, Wallace sipped his morning coffee and focused his gaze on a tall fig tree. And then he saw it: a telltale flash of colour. Within seconds he had abandoned his drink, grabbed his shotgun and charged through the humid forest to the base of the tree. A hunt was on. Many birds came to gorge on the ripe figs that day, but Wallace coveted one species above all others.

The fig tree towered over him. High up in its crown, the bird flapped between branches as it feasted on the figs. Down below, Wallace could spy none of its splendour. A mantle of yellow feathers flowed from the bird's crown to its shoulders. Its throat, cheeks and forehead shone metallic green. A highwayman's mask of black feathers hid its eyes. The only drab thing about it was the chestnut brown of its wings and belly. But this sombre tone was just the background for the bird's bright red display plumes, which peeked out from under each wing, waiting for an occasion to show off. To complete the costume, two long, black, twisted tail feathers trailed behind the bird as it moved.

Down below, gun at the ready, Wallace plotted the beauty's death. But first he had to find it. The crowns of lower trees crowded his view. Leaves shook. Figs fell all around him as the feasting animals knocked them from branches or let them drop half-eaten from their mouths. Wallace spotted his target again. The bird was oblivious. He took aim, raising his shotgun until it was almost vertical. But before Wallace could shoot, the bird had flitted off into the forest. Just as it would do the next day and the day after that.

Wallace later wrote: 'It was only after several days' watching, and one or two misses, that I brought down my bird—a male in the most magnificent plumage.' It was a red bird of paradise. This prize catch was virtually unknown to the wider world. For Wallace, who planned to ship the bird to a trader in London, it meant a payday that would help fund his travel and studies.

Wallace was working his way across the region with his gun, a butterfly net and a mind pregnant with revolutionary ideas about life on Earth. A mix of adventure and misfortune had led him to that fig tree on Waigeo. He was born in 1823, the seventh of nine children, and he grew up in Hertford, a town 20 miles north of London. When Wallace's father was swindled out of a large sum of money, the family fell into poverty. This forced young Alfred out of school and into work at the age of 14. For several years he worked as a land surveyor, and his eyes and mind opened to admit the beauty and mystery of the natural world. But the office work was drudgery he intended to escape, so in 1848, he engineered an adventure to South America.

In the Amazon rainforest Wallace collected thousands of specimens of species that were new to science. Four years later he set sail from Brazil with a great cargo of preserved beetles, butterflies, bird skins and more. Not everything was dead. For the next 28 days, the calls of Wallace's caged monkeys and parrots accompanied the wild songs of the Atlantic's wind and waves. Ahead lay a triumphant return to England.

But it was not to be. On August 6th, the captain came to Wallace's cabin and hit him with a blunt message: 'I'm afraid the ship's on fire.' Wallace and the crew crowded into a pair of leaking lifeboats in time to watch it sink. They drifted for 10 days, sunburnt and sodden, before a passing ship rescued them. Wallace had lost everything but a small case of notes and sketches. His collection had been worth a small fortune—tens of thousands of pounds in today's money. To science, it was priceless.

The disaster would have crushed the spirit of many a traveller. Wallace, however, later called it the 'most fortunate thing' to have happened to him. Without it he would have returned to the Amazon. Instead, in 1854, he headed east to the Malay Archipelago, 'a perfectly virgin country, which hardly any naturalist had then properly explored'. These islands ignited in him ideas that would illuminate the world.

Over Wallace's eight-year stay there, he shot, trapped, bartered for and bought more than 125,000 animal specimens. Glittering in this trove were more than 5,000 species that no scientist had described before, including 212 types of bird. The more Wallace explored the forests, the more he saw patterns in the diversity and distribution of plants and animals. But he was puzzled by the number of distinct yet clearly related species.

The prevalent view among both religious and scientific minds back home in Western Europe was that God had created each species in a fixed form. Wallace, though, was not convinced. He had plenty of questions about the origin of species and how they were so well-suited to the lives they led. Why were species present in some places but not others? Why did the intermediate forms die out, leaving only the distinct species? The forests of the archipelago thrust answers at him. The strangler figs Wallace encountered on every island he visited, and which impressed him more than any other trees, had a part to play. Their roots reached into his mind.

It was in 1858, when a strong fever gripped Wallace, that he hit upon his big idea. He envisioned a 'struggle for existence' among individuals that vary in how well they can survive in their physical and biological surroundings. This variation, thought Wallace, could be the mechanism through which species evolve. It was sacrilegious stuff.

Wallace scribbled down his ideas and sent them by steamboat to his friend and fellow naturalist Charles Darwin. By weird coincidence, Darwin had developed an almost identical idea but had been scared to share it for close to 20 years because it so thoroughly contradicted the Bible. Wallace's letter prompted Darwin to go public. Within weeks, and without Wallace's knowledge, the two men's work had been presented as a joint paper to England's premier gathering of learned naturalists, the Linnean Society of London. The theory of evolution had arrived.

Fifteen months later, Darwin published *On the Origin of Species* and changed the world. While Darwin became a household name, Wallace dallied in Asia, in thrall to its forests. In July 1860, he arrived on Waigeo, where he sought birds of paradise. He later wrote of these creatures that: 'Nature seems to have taken every precaution that these, her choicest treasures, may not lose value by being too easily obtained.' But in Waigeo, Wallace had an ally. He had chosen the site for his hut there in part because it was near the tall fig tree whose leaves would soon feel his gunshot. Six years into his odyssey across the archipelago, he knew he could count on big fig trees to reward his patience.

Wallace's tree was a kind of strangler fig, a group he called 'the most extraordinary trees of the forest'. They lurk in forests throughout the tropics where they can grow into colossal forms. It is not only their size that seizes the attention of anyone who sees them. It's also their shape. They look less like plants than primeval creatures that have

FOREST TAKEOVER:
A strangler fig colonises, embraces and replaces its host tree

frozen in time, their bodies a hulking mass of once-writhing limbs that dangle matted strands of dark hair.

The illusion comes from the tens or even hundreds of false stems that make up a strangler fig's trunk. These stems are in fact roots, for the stranglers don't grow up from the ground but start out in life as seeds that germinate high up another tree. The strangler's roots flow down their host tree, fusing and splitting and fusing again like the molten wax that cools as it courses down a candle.

There's something almost erotic in the way a strangler fig's roots meet, embrace and meld. For this, they feature in a myth among the Wayampi people from the Amazonian rainforest of French Guiana. According to the story, a man learned that his wife had been unfaithful. To take his revenge, he smeared her and her lover with a magical ointment that glued their bodies together. He then threw the adulterers to their deaths from the roof of his house. A giant strangler fig grew out from their bodies, its roots and branches entwined like the lovers' limbs. The Buddha is said to have made a similar comparison, warning that sensual pleasures can leave human bodies 'bent, twisted and split' like the trees that fall victim to a strangler fig's encircling roots.

What Wallace saw in those roots was evolution in action. He noted how strangler figs gain a great advantage over plants whose seeds germinate on a dark forest floor. By contrast, strangler fig seeds germinate high in the canopy and so are immediately immersed in the abundant light they need to grow. After producing their first leaves to capture energy from the sun, and using their first tiny rootlets to absorb nutrients from the hollow in which they find themselves, the strangler seedlings send roots downwards in search of solid ground. When these aerial roots enter the earth, they grow in girth and begin to draw moisture upwards. In a short space of time, these plants have all the advantages of light from the canopy as well as moisture and nutrients from the soil—a very successful short cut. 'Thus', wrote Wallace of the strangler figs, 'we have an actual struggle for life in the vegetable kingdom, not less fatal to the vanquished than the struggles among animals which we can so much more easily observe and understand.'

A strangler fig's roots can create a mesh that encloses the host tree, limiting its ability to grow. It can be a deadly embrace. In some cases, the host tree dies and rots away, leaving behind a hollow core. With the host dead, the basketwork scaffold formed by the strangler fig's roots can still support the growing crown of delicate green leaves and keep the fig alive. Yet even as the stranglers exert their brute power, they show another side to their majesty. They both destroy and sustain.

Once or more each year, they can display as many as a million figs that attract and nourish a stunning variety of birds and mammals. The influence strangler figs have on other species is disproportionate to their numbers, and this makes them important components of many forests. If you want to find wildlife in the tropics, there are few better places to wait than beneath a strangler fig with a ripe crop. Animals will throng there like devotees to a temple, and Wallace knew this.

So Wallace got his red bird of paradise because of the figs. His work—not least because it brought Darwin out of his shell—would set the stage for later scientists to show just how and why fig trees are so special. These trees have provided textbook examples of the beauty and complexity of evolution. This is because the destiny of each fig species is bound to that of various animals, from the partners that pollinate its flowers and disperse its seeds, to the predators that grow fat at the expense of future fig trees.

Ficus species exemplify variation—that key trait Wallace identified as enabling evolution. And so they should. Every day for the past 80 million years, fig trees around the world have been combining their DNA and packing it into trillions upon trillions of seeds. Thus, they have tested a staggering number of genetic combinations, each one an experiment in the struggle for existence. It's a struggle fig trees make look easy.

Not only did these plants survive the cataclysm that saw off the giant dinosaurs and many other species. They flourished. As they spread around the globe, they formed hundreds of new species and became the most varied group of plants on the planet. The side effects have been profound. These plants fed our pre-human ancestors and offered other gifts to the creators of the first great civilisations. Our predecessors rewarded these trees with roles in some of the oldest of our stories.

Judaism, Christianity and Islam all agree, for instance, that figs trees have been part of the human tale since Day One. In the creation story these three religions share, a fig tree was present in the Garden of Eden along with the first people, whom English speakers call Adam and Eve. God had given the couple all they needed and the freedom to do what they liked, but with one proviso—they must not eat the fruit of the Tree of Knowledge of Good and Evil. In Western Europe, we are often told that this fruit was an apple. However, this may simply be because of the Latin bibles that began to spread in the Middle Ages—although they sound different, the Latin nouns for an apple and evil are the same: *malum*. Some Jewish rabbis have concluded that the forbidden fruit was in fact a fig. It was a fig that Michelangelo portrayed when he painted the scene on the ceiling of the Vatican's Sistine Chapel in about 1510.

The story says Eve ignored God's rule and swallowed the fruit. Adam followed suit. They were suddenly aware of their nudity. According to the book of Genesis: 'The eyes of them both were opened, and they knew that they were naked; and they sewed fig leaves together, and made themselves aprons.' A fig tree had come to the rescue, but Adam and Eve's lack of clothes was the least of their problems. God banished them from the Garden of Eden, so preventing them from eating the fruit of the Tree of Life and gaining immortality. This, these three religions agree, is where all of our promise and our problems began.

Like hundreds of millions of people around the world, I heard the story of Adam and Eve at a very young age.

EDEN'S FIGS:
Eve and Adam eat figs from the Tree of Knowledge, in an image first drawn
by a Spanish monk in the year 994

Decades later, and after years of studying the biology of fig trees, the story took on new meaning when I started to learn how fig trees were central to other creation stories too. I learned about Mithra, a Persian deity and 'Judge of Souls'. Some versions of his story say he was born out of a rock beneath a sacred fig tree. Naked and hungry, Mithra hid himself from the howling wind in the branches of the tree, ate figs for his first meal and made himself garments from fig leaves. Then I heard about creation stories from thousands of kilometres away in Africa's Congo Basin. They describe how the first hunter was born from a species of fig tree. Cold and naked, he peeled the tree's bark away and fashioned clothing from it to protect himself.

It may sound uncomfortable but barkcloth is a real thing. Cultures in Africa, Asia and South America have independently worked out how to turn the bark of local fig trees into a malleable material they could wear or write on. People in Uganda still produce cloth from the bark of fig trees. The United Nations has classified the process they use as a 'masterpiece of the oral and intangible heritage of humanity'. Researchers and designers around the world now put the material to new uses, in everything from furniture and high fashion to yachts, cars and aircraft.

Figs feature in other origins too. The Kikuyu people of Kenya have a grove of sacred fig trees at the centre of their creation story. A story from Indonesia describes how two gods formed the first couple from a fig tree, carving horizontal slices of wood to create the woman and vertical slices for the man. A myth told by the Kutia Kondh people in Odisha, India, says the goddess creator Nirantali formed the first human's tongue from the ever-quivering leaf of the sacred fig (*Ficus religiosa*). Another story says Nirantali provided the first people with seeds of another fig species (*Ficus benghalensis*) to plant. The resulting trees provided shade with their thick leaves and, on Nirantali's orders, also fed the first people with their milky latex until grain appeared in the world.

Ficus species don't only feature in creation stories. They also represent gods and serve as abodes for spirits. They can be symbols of divine sustenance or ethereal bridges that link heaven and earth. In East Africa, Maasai people tell stories that say when the earth and the sky became separated, all that connected them was a sacred fig tree. It was via this tree's aerial roots that the Maasai god provided cattle to the people. Among the Akan people of Ghana, on the other side of the continent, one of the first traditional duties of the ohemaa—a female ruler also called a queen-mother—was to create a sacred place by planting a fig tree. Figs are also cosmic trees in Candomblé, a religion that originated in northeastern Brazil when slaves taken there from West Africa found fig trees just as impressive as the sacred figs from which they had been separated.

In Hong Kong, where people say *Ficus microcarpa* fig trees are home to spirits, two of these trees have become famous as 'wishing trees'. For many years, people would come to throw oranges into the trees' crowns. They had written their wishes on strips of crimson paper, which they attached to the oranges with string. If a wishing tree's branch caught hold of the missile, their desires could dangle and be blown by a breeze up to heaven. But, in 2005, when the weight of all those wishes caused a branch to break off and injure a man and a child, the government banned the practice.

Far away on the Indonesian island Sumatra, the Batak people have a fig for their 'world tree'. This is a mythical *Ficus benjamina*, the same species whose leaves I dusted as a child. The Batak say their tree grows among the stars and that its roots reach down to earth. Mortals can clamber up them to reach heaven. On the island of Borneo, the Iban and other indigenous peoples traditionally prohibit the cutting of strangler figs because spirits dwell among their roots. In Myanmar, for a millennium before Buddhism became the main religion, people worshipped spirits called 'nats' including Nyaung Bin, an old man who lives in a fig tree. In the Philippines, fig trees are said to be home to supernatural beings such as giant tree demons, goblins and the half-human, half-horse tikbalang.

To the northeast, on the Japanese island of Okinawa, folk stories feature short, red-haired spirits called kijimuna that inhabit fig trees. Far to the south, in Timor-Leste, the Sun god Upulevo is said to have settled on a fig tree to impregnate his wife, the Mother Earth. In the Sepik River area of Papua New Guinea, people consider fig trees to be an abode of crocodile spirits. Meanwhile, in Australia, aboriginal communities warn of the yara-ma-yha-who, a blood-sucking manlike creature that lives in fig trees and preys on unwary travellers. And on the Pacific Ocean island of Guam, thousands of kilometres from the nearest sizeable land mass, ancestral spirits called taotaomonas are said to live among the roots of fig trees.

These are just snapshots. There are bigger, better and much more profound stories to tell. Search for these stories and you will find them across a great swathe of the planet. They are ancient stories. They come from a time when nature formed the foundation of faith, and when science had yet to ask its first questions. What science has since shown is that there are good reasons for the preponderance of fig trees in diverse religions. This points to lessons for our modern world, and to potential bridges between sceptical and romantic minds.

Wallace saw no contradiction between spirituality and science. In his first major scientific paper on evolution, he found space to write of the wonder he felt when he set foot in the great forests of Borneo. 'When, for the first time, the traveller wanders in these primeval forests, he can scarcely fail to experience sensations of awe. . . . There is a vastness, a solemnity, a gloom, a sense of solitude and of human insignificance which for a time overwhelm him.'

Psychologists Dacher Keltner and Jonathan Haidt have suggested that awe is something we evolved to feel. Such sensations may have given our ancestors an advantage in their struggle for existence, they say—by making them respect more powerful entities or pay attention to their environment, for instance. If so, perhaps this helps explain why *Ficus* species have become embedded in so many diverse cultures. Tropical forests are indeed awesome, and giant fig trees are among the most awesome things within them.

Their power and fertility demand the attention and respect of every human eye that sees them.

My time to feel the awe Wallace described came in 1994, when I followed Charles Darwin's advice that: 'Nothing can be more improving to a young naturalist than a journey to distant countries.' I went to Sri Lanka. There I encountered *Ficus religiosa*. This species of strangler fig includes one of the most famous trees in the world, a tree that has travelled thousands of kilometres, has been made a king and has helped spread a philosophy of peace.

THREE

A Long Seduction

The Buddhist monk's robe sang out loud saffron over the rainforest's muffled tones of brown and green and grey. The monk walked fast. He was a tall tree's length away when he paused on the path and turned to face me and my friends. Sunlight cut through the forest canopy and shone hard off his shaved head. With one hand, he raised aloft a black umbrella. With the other he beckoned: come, follow.

It was 1994—four years before I found myself hanging from that ladder in Borneo. I was in the Udawattakele Nature Reserve in deliciously named Kandy, a historic city nestled in the highlands at Sri Lanka's heart. I had gone there with three university friends during our summer break, and for me, the biology undergraduate, it was paradise. I had blown my student loan on the plane tickets and I had no regrets. It was my first time in a tropical forest. It changed me.

Overhead, golden-backed woodpeckers shot like painted

arrows between the trees. Other animals chuckled, cooed and coughed from hideaways high above me. Butterflies fluttered by and flirted in the sunbeams. Dragonflies, big as teaspoons, patrolled coffee-coloured ponds where turtles gazed lazy-eyed, secure on semi-submerged logs.

But all these things seemed insignificant in the presence of the forest itself. It hugged all it contained in a humid, humming gloom. The outside world seemed remote now, the sun an intruder. It sneaked peeks through breaks in the leaves but caught only glimpses of the life that throbbed below. The trees towered over us, viscerally alive yet so alien to our animal ways. Their breath sweetened the air we inhaled. It is hard to explain, but I could feel the concentration of life around me, as if its great density there had somehow reached into me physically. What struck me was the neutrality of that force. There was no malice or love there, just existence.

The monk smiled like the Mona Lisa and we decided to follow. He strode in silence and led us through the forest to a clearing and a rock overhang that formed a shallow cave. The cave had been enclosed by a whitewashed wall in whose centre was a wooden door. The monk unlocked the door and disappeared behind it, only to return seconds later with a broom. He swept the flat area outside the door, scattered some scraps of bread there, then sat down on a large slab of rock. He had yet to say a word.

The sun blazed and all around the forest crowded in on that little clearing. Within a minute the closest trees began to shake. Leafy branches crashed against one another. It was a mob of monkeys, more than a dozen of them. They poured down the tree trunks to feast at our feet on the torn bread. The monk watched in silence and smiled.

The monkeys were toque macaques, a species that lives only in Sri Lanka. Figs are among their favourite foods and it is thanks in part to the Buddha that they have a steady supply. For the monk, a fig tree meant much more than mere food. I realised this when he broke his silence to invite us through the wooden door. Beyond it was a small cave whose cool walls shone white with paint. He said monks had lived and meditated in the cave for the past 2,000 years. There was no furniture, just a small cushion, some books and a plastic-wrapped painting of the Buddha, sitting cross-legged and deep in meditation beneath a fig tree. The Buddha had wandered for six years before he found that tree. When he found it, he also found enlightenment.

The man in the picture hadn't always been at peace. According to Buddhist lore, he was born into royalty around 563 BCE in what is now Nepal. The young prince, whose name was Siddhārtha Gautama, grew up insulated from the hardships of life. But he grew unhappy after he witnessed sickness and death, poverty and the decay that time inflicts on once-strong bodies. These senseless things disturbed him. At the age of 29 he left his home and his riches, his wife and his child, and went off to wander the world in search of meaning.

Gautama studied under wise men and lived for some time in a forest as an ascetic, but still he found no answers. On he trekked. Six years after he left home, he arrived at a forest near the city of Gaya in what is now the Indian state of Bihar. There he found a fig tree. When he sat beneath it to meditate, he pledged not to leave until he had liberated his mind.

The tree belonged to a species scientists today call *Ficus religiosa*—the sacred fig. This species grows up to 30 metres tall and has smooth grey bark and small red figs. Its hand-sized, heart-shaped leaves are shiny and stiff, with long pointed tips and long slender stalks. When the wind blows, the leaves tap against each other and create a sound like the wing beats of thousands of tiny birds. This fluttering filled Gautama's ears as he tried to fathom the meaning of the universe.

After six days and six nights he achieved his goal and attained enlightenment. He had found an explanation for human suffering and a way to end it. He had become the Buddha. Some stories say he was so grateful to the fig tree that he gazed at it with motionless eyes for a full week. He stared at forest royalty.

Ficus religiosa is one of the strangler figs. It does best when it grows from a seed that has landed not on the ground but in a hollow on a tall tree. When its aerial roots reach the ground, they thicken into strong scaffolds. When its figs are red and ripe, they won't last long before colourful birds fly in to feed. Rufous treepies and golden orioles, rosy starlings and yellow-footed green pigeons are among the species that love these figs. Monkeys might swing along too, but at night large fruit bats called flying foxes often take over. Chital deer and nilgai antelopes will forage on any figs that fall to the ground. Some of these creatures depend on such fig feasts when other fruits are scarce, and most of them can disperse the fig tree's seeds. They help ensure the species they feed upon survives.

Unlike other strangler figs, which use their hosts only for support, a *Ficus religiosa* can carve its way into its host's wood, splitting it apart. These fig trees can live for hundreds of years, long after all trace of their host has gone. By now they stand free. Their robust roots have merged to form a stout trunk in whose deep-fluted hollows a wanderer can sit to shelter and contemplate the world. Siddhārtha Gautama is said to have done just that. His purported choice of tree would shift the fortunes of the entire species. But it was not the first time the fate of *Ficus religiosa* had entwined with that of the people who walked beneath it. This species had insinuated itself into human affairs thousands of years earlier.

Ficus religiosa had long played the game of life in forests where no human trod. Its struggle for existence was marked by conflict with seed-destroying animals and partnerships with pollinators and seed dispersers. Its life was a balancing act. Then, tens of thousands of year ago, along came a new player who tipped the scales. It was our species. We were few in number at first but would leave deep footprints. We had developed a way to loosen the shackles of evolution. We had culture. Our ideas and technologies could spread faster than our genes.

With our big brains and grasping hands, our axes and our fire, we would have a disproportionate impact on the species that surrounded us. Against the forests and *Ficus religiosa*, the dice were now loaded. Unlike most forest species though, *Ficus religiosa* was able to gain by becoming part of that which sets humans apart: our culture.

It began early. Those first people found in *Ficus religiosa* giants among trees. Unstoppable, they grow into hulking forms. Yet each has its own distinct shape and character. They would have been landmarks to the wandering bands of forest people. They were also sources of food. Each periodic outburst of fruiting drew dozens of wild animal species. Those first people to encounter *Ficus religiosa* would have joined the feast but also hunted the animals they dined alongside. Over thousands of years, as the human population grew and began to settle and farm the land, new cultures developed. The value of *Ficus religiosa* would shift from the material to the symbolic and the sacred.

By 3300 BCE, the Harappan people of the Indus and Saraswati valleys of what are now Pakistan and India were laying the foundations of a great civilisation. They cleared forest to plant crops and build houses. Their cities were the most advanced settlements of the time, with the world's first sanitation systems. The Harappan people had transformed a wild landscape into the pinnacle of urban planning. *Ficus religiosa* rode out these great changes. The tree was special to these people.

The earliest depictions of any tree in South Asian art or literature are the images of *Ficus religiosa* the Harappan people left more than 4,000 years ago on small soapstone seals. These are stone stamps that a merchant or administrator might have used to mark clay tags on documents or packages. Some seals show a human figure beneath an arch of *Ficus religiosa* leaves. Others depict a person with a leafy branch from the tree on their head. One intriguing seal shows a man kneeling before a *Ficus religiosa* tree. Within the tree stands a figure, their arms bedecked with bangles, their hair in a single long plait. Alongside the kneeling man is what appears to be a severed human head. Some scholars think the scene represents aritual sacrifice to a spirit or deity who resides in the fig tree.

The image offers a tantalising glimpse into a mysterious people whose writing remains undeciphered and whose civilisation fell into decline when calamity struck. Around 3,500 years ago, the rivers this culture depended on changed course and the people abandoned their cities. For their favourite trees, the future was uncertain. But there were strangers on the horizon, new minds to seduce. The changes began around 1500–1400 BCE, when nomadic herders moved into the Indus Valley. The migrants would come to dominate the landscape, in time settling and farming the land. They brought their own gods and stories, but as they mixed with the people already there, something new was born: the Vedic culture.

The surviving Vedic hymns reveal just how much this new force valued *Ficus religiosa*. They tell how the people used the tree's wood in important rituals to ignite fire and prepare a hallucinogenic drink. *Ficus religiosa* was central to these rituals that helped bind the new culture together. The trees were also early pharmacies. People in India today

A FIGURE IN A FIG TREE:
An ancient Indus seal depicts what may be a sacrifice to a deity in a fig tree

A Long Seduction

use medicines made from this tree's bark, leaves or roots to treat dozens of conditions. Some of these remedies date back thousands of years to the Vedic culture.

The Vedic people called *Ficus religiosa* an 'abode of the gods' and home of the 'universe's mighty keeper'. They adopted the species as a symbol of strength, a destroyer of enemies. A Vedic prayer directed to the fig tree includes this vivid couplet: 'As you climb up the trees and render them subordinate, so split in two the head of my enemy and overcome him!' *Ficus religiosa* had once again become cemented into the dominant culture. Greater roles were to come.

As the Vedic culture spread east into the dense forests of the Indo-Gangetic Plain, it encountered local cults who believed spirits dwelt in *Ficus religiosa* trees. The Vedic culture assimilated these people along with their stories. The blend of cultures gave rise to Hinduism, which elevated *Ficus religiosa* to new heights. The species symbolises three core Hindu deities—the roots representing Brahma the Creator, the leaves Shiva the Destroyer and the trunk Vishnu the Preserver. The many other Hindu gods associated with this kind of fig tree include Krishna, Shani, Hanuman and Lakshmi.

Ficus religiosa has been described as both a tree of knowledge and tree of life, a place people can go to pray for fertility and longevity. This species would become sacred to every Hindu caste, a tree to both protect and worship. To harm one became a deadly sin, so these trees were free to grow into giant forms. Where better for

Siddhārtha Gautama to seek enlightenment? The species had survived thousands of years in the company of humans. It had thrived, but thanks to the Buddha's decision to meditate beneath one, its fate took a new turn.

After the Buddha died in c.483 BCE, his followers began to flock to the fig tree, which they treated as a living embodiment of the Buddha himself. The tree became known as the bodhi tree, or tree of enlightenment, from the Sanskrit word *bodhi*, which means awakened or knowing. It would become the most famous tree in the world. Few of the pilgrims would have such an impact on the future of Buddhism—or on the fig tree—as a man called Ashoka who lived from 304–232 BCE and, as Emperor Ashoka the Great, became the third in his line to rule the vast Mauryan Empire.

Accounts of Ashoka's early life paint him as a monster. He is said to have murdered 99 half-brothers, burnt alive 500 women he kept in a harem, and executed 500 of his ministers. He ruled over much of what is now India, as well as parts of Afghanistan, Bangladesh and Pakistan. But something was missing: the Kalinga Kingdom, which neither his father nor grandfather had conquered. In 261 BCE, Ashoka poured in 400,000 soldiers. They outnumbered those defending Kalinga by more than six to one.

For days sounds of violence filled the air, and at the end of it all, the Daya River ran red with blood. More than 100,000 civilians lay dead. Ashoka captured and deported thousands more who survived. Kalinga was broken. Ashoka had achieved what none of his ancestors had. But he had

36

won like a tiger against a cat. A more difficult conquest was soon to come — the conquest of himself. The turning point came when Ashoka surveyed the carnage he had wreaked.

Corpses buzzed with flies. Vultures circled overhead. The stink of burnt and rotting flesh filled Ashoka's nostrils. The wails of orphaned children and widowed women assailed his ears. The weight of responsibility was crushing. Ashoka renounced violence and said in future he would conquer by kindness. Ashoka the Wicked had become Ashoka the Righteous. He embraced Buddhism and made it the official faith of his empire. For *Ficus religiosa* this would mean a passport to world travel.

The change of fortune began in about 250 BCE when Ashoka visited the fig tree where the Buddha had attained enlightenment nearly three centuries earlier. Ashoka consecrated the site, built a shrine there and held a festival every year in honour of the tree. For hundreds of years before any image of the Buddha appeared, it was the fig tree surrounded by adoring devotees that sculptors portrayed at Buddhist temples.

In time, the original bodhi tree died, but Buddhists say it lives on elsewhere because of how Ashoka spread his message of peace to another recent convert, the Sri Lankan King Devanampiya Tissa. And what better way than by sending him a branch of the bodhi tree itself? Never before nor since has a plant travelled in such style. Its journey has become steeped in legend as described in the *Mahavamsa*, or 'Great Chronicle', an epic Pali-language poem that recounts 900 years of Sri Lankan history.

TREE OF ENLIGHTENMENT:
Winged spirits and earth-bound devotees at the Buddha's *Ficus religiosa*,
from a first century BCE sculpture at Sanchi, India

The *Mahavamsa* says Ashoka had the branch planted in
a vase of solid gold that was eight-fingers thick and had a
rim the size of a young elephant's trunk. Ashoka bestowed
kingship upon the plant and appointed a diverse retinue,
dozens strong, to accompany him on the branch's journey
to the sea. The company included nobles and cowherds,
potters and weavers. Royal maidens watered the branch in
public ceremonies at key stages of the journey. The branch
and its entourage first voyaged by ship down the River
Ganges to the Bay of Bengal. Here, Ashoka's daughter,
Saṅghamittā, and his son, Mahinda, took the bodhi tree's
branch aboard a seafaring ship. As the ship sailed away,
Ashoka—once the hard man of the subcontinent—stood on
the shore and shed tears.

When the ship arrived in northern Sri Lanka, Devanampiya Tissa was there to meet it. Sixteen nobles strode out
into the warm sea with the King. They walked on until the
water lapped around their necks so they could collect the
golden vase and carry it ashore. After a long journey inland,
Devanampiya Tissa planted the branch in his capital Anuradhapura. He employed archers to protect the tree from
foraging monkeys, relatives of the fig-loving monkeys I
saw in the Udawattakele forest that day in 1994.

Visit Anuradhapura today and you will see a giant *Ficus
religiosa* that Buddhists say is the same one Devanampiya
Tissa planted, making it the world's oldest living tree with
a known year of planting. Sceptics say it is more likely that
today's tree is a descendant of the original. What's certain
is that the fig tree began life a long time ago. It is vast now,

its branches thick and strong. They reach for the sky and support a wide crown of leaves, which pilgrims treasure as souvenirs when they fall to the ground. Buddhists come from all over the world to see the tree and make offerings of rice and lotus flowers. The air is perfumed with incense and coconut oil. Colourful prayer flags festoon the golden rails that surround the tree and protect it from the thousands of visitors who flock to see it each year.

This tree, the Sri Maha Bodhi, has itself helped spread the Buddha's philosophy. Saplings that sprang up near it were sent to various parts of Sri Lanka. Someone humped one up and down many hills to Kandy where a monk planted it in the Udawattakele forest in which I walked centuries later. Today clones grown from cuttings of the Sri Maha Bodhi tree can be found at Buddhist temples throughout Sri Lanka and in Buddhist communities the world over.

In March 2011, some of the pomp and ceremony that surrounded the tree's arrival in Sri Lanka burst out again. Monks took a cutting—little more than a slender branch with around 20 leaves—from the Sri Maha Bodhi and placed it in a gilded casket so it could journey back to Bodh Gaya, the site of Buddha's enlightenment in India. The circle was complete—a round trip of 3,600 kilometres.

When the original branch left India for Sri Lanka, it left a land of thick forests and bullock carts. The branch that returned now grows in a country with skyscrapers and a space programme. Much has changed, but much has stayed the same. Today millions of Hindus, Buddhists and Jains there consider *Ficus religiosa* to be not just a tree but an embodiment of divine power that can bring benefits to those who worship it. The species features in an array of rituals that include prayers for fertility, marital bliss, health, wealth and good luck.

Among the most famous of people reported to have had a close encounter with this tree species in modern times is Aishwarya Rai, the Bollywood superstar and former Miss World. According to Hindu astrology, she was born a 'manglik'. This means the positions of the planets at the time of her birth carried an omen, a prophecy of misfortune. This could manifest through marital woes, such as fighting, or even the untimely death of a partner. When Rai and actor Abhishek Bachchan decided to marry, this put him and their relationship in the firing line. So, according to media reports, Rai followed an ancient tradition and first married a fig tree, a *Ficus religiosa*, which would absorb all of the bad luck and leave her free to then marry her human love. Her creative solution was part of a chain of events that has tied humanity to this special fig tree for more than 5,000 years, since the time of the Indus Valley Civilisation and long before that. It is a chain upon which the Buddha's enlightenment beneath a *Ficus religiosa* is just one link.

When I looked at that painting of the Buddha beneath the tree in that forest cave in 1994, I still had no idea about the secret to the fig trees' success. But I can draw a direct line between that encounter and my journey into the story of *Ficus*. In that moment of golden peace and the company of monkeys and a high canopy of tall trees, something inside me clicked. The forest had touched me. If this was awe, I wanted more. Fig trees would be my gateway.

Ficus religiosa is not alone. It is one of more than 750 fig species, each with its own story and its own role in our story. The journey to understand this great diversity of species began more than 2,300 years ago in Greece. While fig trees played many important roles in the development of Eastern philosophy, the early Western philosophers had figs of their own to ponder. And as the first followers of Buddhism began to spread the faith, far to the west these other fig trees were present at the birth of biology, the science that would reveal why *Ficus religiosa* ever mattered enough to make people revere it in the first place.

FOUR

Banyans and the Birth of Botany

Sykeus was big. Sykeus was strong. But Sykeus needed his mummy. He was on the wrong side of a one-sided war, and it was all her fault. His mother was Gaia, the Earth goddess of ancient Greece. She had compelled Sykeus and his siblings to rise up and overthrow the gods on Mount Olympus. But her offspring were hopelessly outmatched.

Gaia watched helpless as Zeus and his fellow Olympians slew them one by one. They were clubbed to death, crushed by rocks, torched or flayed alive. Sykeus was next. Zeus pursued him, flinging bolts of lightning at him as he fled. Just when Sykeus could run no more Gaia opened up her chest, took him inside and transformed him into the first fig tree. And that is why there are fig trees in the world.

Oh no, argues another ancient Greek account, it is the fertility goddess Demeter we should thank. She gave figs to humanity to repay the hospitality of King Phytalus after he gave her shelter when she sought her lost daughter,

Persephone. But no, insists a third story, that's simply not true. It was Dionysus, the Greek god of wine, who found the first fig tree. Tasty as these tales were, they could not sate the intellectual hunger of a philosopher called Theophrastus who lived from c.371–287 BCE. Theophrastus hankered for truth.

Theophrastus is little known today, though his impact was immense and when he died Athens mourned en masse. He studied and wrote about history and philosophy, poetry and grammar, politics and religion. But his greatest legacy comes from his scientific enquiries into the lives of plants. Not for nothing is he considered the founder of botany. For more than 1,500 years after his death, nobody made a greater contribution to the field. He was the first in a long line of men and women who, by peering at figs, have peeled away layers of mystery and exposed reality to the light of their reason. But nearly 2,400 years after Theophrastus began to name and describe the world's fig species, the task is not yet complete.

The fig species Theophrastus knew best was *Ficus carica*, a tree with pale grey bark, unmistakeable lobed leaves and plump perfumed figs that ripen purple, green or black. This species, the 'edible fig', had been an important source of food in Greece for thousands of years. By the time of Theophrastus, the citizens of Athens were known as *philosykos*—literally 'fig lovers'. His contemporary, the poet Alexis, wrote of 'that god-given inheritance of our mother country, darling of my heart, a dried fig'. These figs were like treasures. Theophrastus noted how the climate and soil

44

and strange little insects combined to decide the quality of fig crops. But he was just scratching the surface of one of biology's most fascinating stories.

As Theophrastus focused on the *Ficus carica* trees all around him, he was unaware of hundreds of other fig species, each with its own way of solving life's challenges. He didn't know the interiors of figs are scenes where great dramas play out daily, with consequences for thousands of off-stage actors from the plant and animal worlds. Patchy though his knowledge was, Theophrastus still came to learn about the most astounding fig species of them all. It was thanks to the unrelenting drive of one of the most powerful people to have lived: Alexander the Great.

In 326 BCE, after conquering swathes of what are now Greece, Turkey, Syria, Lebanon, Israel, Egypt, Iraq, Iran, Afghanistan, Uzbekistan, Tajikistan and Pakistan, Alexander and his army reached India. Alexander valued knowledge about the nature of the lands he occupied so had brought along naturalists to collect specimens and report on the fauna and flora. For these explorers, everything was new. They met monkeys, massive snakes and beautiful birds such as parrots and peacocks, none of which they had ever before seen. The plant life was no less exotic, but one species seized their imaginations like no other. West had met East and found the giant among fig trees, the banyan (*Ficus benghalensis*).

At first, it must have seemed like a forest. Hundreds of trees held aloft a vast green canopy that cast deep cool shade. When a group of Alexander's men stepped under that leafy umbrella, they realised it was in fact a single tree, though one with hundreds of 'trunks' that propped up its biggest branches. Alexander's admiral, Nearchus, said ten thousand people could have sheltered under that banyan.

The tree was old. Many years earlier, another tree had occupied that spot. Its fate shifted when a bird, or perhaps a bat or monkey, passed by having fed earlier on ripe *Ficus benghalensis* figs. The animal pooped on the tree and condemned it to a slow death by smothering. The animal's droppings had delivered a banyan seed to a moist nook. Within weeks, the fig seed had split open. It sent up a firm stalk with a collar of two tiny green leaves. It sent down tiny roots that hugged the host tree as they stretched earthwards in search of soil. In time these roots would expand and enlace. They would encase the host tree and erase all trace of it.

As the banyan grew, its branches also sent out roots. They dangled like strands of unkempt hair. When they reached the ground, these roots grew thick and woody and merged to form what looked like new tree trunks. The massive branches reached ever outwards, sending down yet more and more prop roots. These false trunks formed increasingly wide circles around the banyan's core, enclosing it in nested cloisters.

There is little to stop a banyan expanding. The biggest one on record is said to have begun life in 1434 at the spot where a woman called Thimmamma died when she threw herself onto her husband's funeral pyre. That tree, in Andhra Pradesh, now covers two hectares. Twenty thousand people can shelter beneath its crown.

BENEATH A BANYAN:
A single *Ficus benghalensis* can resemble a small forest thanks to the
false trunks its pillar roots form

All this from a seed that is just a couple of millimetres in length. Crack one open with your thumbnail and you won't find much inside, yet the genetic material within has the power to create a tree vast enough to resemble a small forest. Long before Alexander arrived in India, Hindu sages employed this paradox in a parable, which used the imperceptible power within a banyan seed as a parallel of Atman, the invisible essence Hindus say permeates and sustains the universe and all it contains.

The sheer size and weird form of India's banyans amazed Alexander and his men, but local people saw them as much more than impressive trees. The banyans had been part of the cultural fabric for thousands of years. Settlements had grown up around these trees. To bodies, they provided shelter, food and medicines. To minds, these awesome structures formed bridges to the supernatural. Gods and spirits moved among the banyan's leaves and pillar roots. By 500 BCE, Hindu texts described a cosmic 'world tree', a banyan that grew upside down with its roots in the heavens, and its trunk and branches extending to earth to bring blessings to humankind.

The banyan became a potent symbol of fertility, life and resurrection. It features in stories of the Maha Pralaya, a periodic death and rebirth of the universe, when everything that exists dissolves into a ceaseless sea. One version of the story says an 'undying' banyan is the only thing to survive the deluge. Another says that to ride the sea's currents, the god Vishnu assumes the form of a baby, lying on his back on a raft formed of a banyan leaf. With one breath the baby swallows all that surrounds it, taking the turbulent universe into the safety of his stomach before exhaling it into fresh existence.

Another Hindu story from more than 2,500 years ago tells how a woman called Savriti convinced Yama, the god of death, to resurrect her husband who had died beneath a banyan tree. Today, married women in north India emulate Savitri's devotion in an annual ceremony in which they tie coloured thread around a banyan tree while praying for the wellbeing of their husbands.

These symbols of love and life became agents of death after the British arrived in India and began to subjugate the local population. They defiled many sacred banyans by using them as gallows to execute rebels who resisted their rule. By the 1850s, there had been multiple occasions when they hanged over a hundred men to death from a single banyan tree.

Alexander's army were more respectful of these trees. They had come from a land where people revered figs, thought them to be divine, used them as food and medicine and symbols of fertility, and said spirits dwelt among them. Thousands of kilometres away, they found people who held remarkably similar beliefs about utterly different fig trees. The parallels are a testament to the depth *Ficus* species have implanted themselves into human culture.

In time, the Greek explorers' tales of banyan trees reached Theophrastus, who wrote: 'The Indian land has its so-called "fig tree", which drops its roots from its branches every year . . . the fruit is very small, only as large as a

A BANYAN'S BLESSINGS:
In an ancient Hindu tradition married women tie threads around a banyan tree
and pray for their husbands' wellbeing

chickpea, and it resembles a fig.' With these words he joined two distant dots, which his intellectual heirs would reveal to be part of a glittering constellation of hundreds of fig species. And so began the long journey to identify them all. It's a journey whose end biologists say seems near but, more than 2,000 years later, is still somewhere around the next corner.

Theophrastus's words echoed through time to influence the eighteenth-century Swedish botanist Carl Linnaeus who developed the system scientists use to name species and gather them into related groups. It was Linnaeus who gave fig trees the formal scientific name *Ficus*. He was using a ready-made word in Latin, the language of Rome, a city whose origin story happens to feature a fig tree. The legend says Rome was founded by Romulus and Remus, twins rescued from drowning in the River Tiber by the fig tree's roots. Under the convention Linnacus developed, each species receives a two-part name: a noun followed by an adjective. The first part is shared by each member of a genus (e.g. *Ficus*) of closely related species. The second part is unique to each individual species.

Linnaeus gave the name *Ficus carica* to the common domesticated fig species after Caria—a region of ancient Anatolia in what is now Turkey. The scientific name he gave the other fig of my childhood, *Ficus benjamina*, has a more convoluted origin. Cut the tree and white latex will bleed out. Various other species also produce this particular kind of sticky fluid, which people have used for centuries to make perfumes, incense, medicines and other products.

This substance is known as gum benzoin, from an Italian interpretation of a Javan word that is Arabic in origin. English tongues mangled the word some more to form 'gum benjamin'. So over time, the benzoin trees ended up being called benjamin trees, hence the benjamin fig (*Ficus benjamina*). I prefer its better-known name—the weeping fig— which it got because, when it sheds its leaves, they fall like green tears to the ground.

Linnaeus mentioned just seven species of *Ficus*. While that is more than double the number Theophrastus wrote about two millennia earlier, it is still less than one per cent of the world's fig species. Since Linnaeus, scientists have identified about 750 *Ficus* species and are still finding new ones. We know about much of this diversity because figs captured the imagination of a man called Edred John Henry Corner, a titan of tropical botany who was as flawed as he was brilliant. What Corner learned over decades studying figs shows just how wrong the Greek myths were. The fig tree wasn't a gift from Demeter or Dionysus. Gaia didn't form the first *Ficus* from Sykeus, her son. No, the fig trees came from forests far away, and they arose long before the first people gave voices to the first gods.

FIVE

Botanical Monkeys

It's 1937, and in a dense rainforest in what is now Malaysia, a tall Englishman with a short temper is shouting at a giant of a tree. It is EJH Corner, his white shirt tucked into khaki shorts, his white socks reaching near to his knees. He looks like an overgrown boy scout except a smoking pipe hangs from the corner of his mouth, angled like a shotgun waiting to be loaded. '*Ambil itu,*' he shouts upwards. '. . . *Ambil itu!*'

More than 40 metres above him, the rainforest canopy abounds with life. A monkey scampers along one of the tree's uppermost limbs, its eyes and mind alert for something tasty to eat. Large figs adorn a creeper that hugs the branch, but the figs are green and unripe: no good to the monkey yet. Perhaps it will find a lizard to catch, or a juicy cicada to stuff into a cheek pouch. Then something clicks in the monkey's mind. It remembers what '*ambil itu*' means . . . 'bring that.'

The monkey tears loose one of the hard figs and hurls it at Corner. And then another. And another. The monkey is now a flash of brown fur, grabbing more figs and sending them hurtling down. Corner runs for cover from what he will later liken to a barrage of hand grenades. When the violence ends, he returns to the scene. Scattered in the leaf litter are about fifty figs of a kind he has never before seen. He has struck botanical gold.

Corner had left England eight years earlier, aged just 23, to become assistant director of the Singapore Botanic Garden. It was a plum job, but Corner despised the colonial social scene, with its golf games, drinks and vacuous chatter. Whenever he could, he escaped across the causeway linking Singapore to Malaya, where he seized '. . . the unparalleled opportunity to explore primeval forest at every step.' The forests were tall, dense and rich. They were also under threat from logging and the rapid spread of rubber and oil palm plantations. Corner vowed to study them before they fell.

'To be in the jungle is a biological consummation', he wrote in 1930. 'To stumble among the riot of enormous trees and to cut a path through the tangle of creepers which knit the life of the rainforest into one gigantic web, is like a dream.' And in that dream grew the strange plants that would occupy Corner's mind for the rest of his days: the *Ficus*.

They were everywhere. Corner encountered giant strangler figs whose sinuous aerial roots resembled masses of snakes. He found trees that dangled ropes of figs from their trunks. He found climbing species of *Ficus* that hauled themselves up big forest trees and epiphytic *Ficus* species that lived high in the canopy and never needed to send down roots.

Just as these plants grew in different ways, so did their figs vary. Some were smaller than a pea, others as big as a tennis ball. They ripened red, orange or green, purple, brown or black. Some figs were smooth, others hairy. Among the strangest species was a tree called *Ficus treubii*. Its figs are the off-white of a smoker's teeth. They grow not on leafy branches but on ground level runners which scramble over and under the soil. This species buries its figs like dirty secrets.

'By themselves the figs could build a forest,' Corner wrote. It was a comment on the large numbers of very different *Ficus* species that can coexist in one place. As biologist Rhett Harrison has shown, many tropical forests around the world have more *Ficus* species than species of any other genus of plants. For Corner, this meant plenty of opportunities to encounter fig species he had never before seen. As he tramped through the forests around him, he discovered and named dozens of new *Ficus* species. In their great variety, he saw an opportunity to understand how the world's plants had evolved to display such diverse forms as tall trees, climbers and epiphytes.

But Corner had a problem. Evolution had run wild. Whichever way he glanced, his eyes met hundreds of different tree species. To tell species apart, Corner needed to examine their leaves and flowers, their fruits and seeds,

ALL SHAPES AND SIZES:
Figs of ten *Ficus* species collected by the author on a single day in Borneo

often for microscopic differences. But the trees soared to 50, 60 or even 80 metres in height and kept their botanical clues far beyond his reach. Until, that is, he hit upon his idea of 'a little hand in the canopy'.

Corner had seen trained monkeys called pig-tailed macaques climbing tall palms to harvest coconuts. He reasoned that if they could do that, they could collect other things too. So he bought a monkey and began to train it to follow new commands. Soon he had four of what he called his 'botanical monkeys'. He joked that they were the first primates to become civil servants, as the then Straits Government paid an annual allowance to provide each with a collar and lead and a supply of rice, bananas and raw eggs. Merah, the monkey that bombarded Corner with unripe figs, managed to pluck samples from more than 350 plant species in just six months. Another monkey Corner called Puteh was the star, able to understand 24 Malay words. But as Puteh matured, he reminded Corner that he was a sentient being with a mind of his own.

One day Corner was in his garden when Puteh charged at him with 'open jaws and slobbering fangs'. Corner raised his right arm to protect his face and neck just in time for Puteh to sink his teeth into the arm and clamp his jaws shut. Only by tearing his arm from the monkey's mouth could Corner escape. The bite was severe. It severed a nerve. A long slab of flesh hung down from its bone. After surgery and a week in hospital, Corner returned home with his arm in sling and advice not to use it for four months. He banished Puteh to a cage. The monkey would botanise no more.

The injury was the least of Corner's problems. It was 1941—the world was at war. Singapore was Britain's strategic jewel in southeast Asia and Japan wanted it. Its forces were advancing fast. Corner reckoned that if the fighting reached Singapore, Japanese soldiers or local looters would destroy the botanical garden's biological collections and the cultural treasures in the Raffles Museum.

In February 1942, the attack began. For a few days Japan bombarded Singapore's defences, then came the ground assault. Corner knew which way the wind was blowing. He liberated the last of his botanical monkeys but could not free Puteh for fear he would attack somebody. With deep regret, Corner shot Puteh dead. 'I did not then know', he later wrote, 'that I killed the one who in all probability saved my life.'

Five days later Britain surrendered. What followed was three and a half years of brutal occupation. Japan summarily executed many ethnic Chinese civilians and imprisoned the British, Indian and Australian military men, working many to their deaths in forced labour camps. Had Puteh the monkey not put Corner's arm out of use, Corner would have remained conscripted into the Singapore Volunteer Force. Instead he was invalided out and so avoided becoming a prisoner of war.

Corner should still have been interned in a civilian camp, but he avoided this grim fate too. Instead, he and two of his colleagues were placed under house arrest at the botanical gardens, where they continued to work. As Japan committed atrocities all about him, Corner spent days staring down a microscope at fig seeds. Many of those languishing in the camps branded him a collaborator, a label he didn't shake off before he died more than 50 years later. In truth, it was a strange act of scientific diplomacy that left Corner free while his compatriots languished in the camps. It was driven by Corner's ability to see the bigger picture and his belief that science was all that mattered.

Before Britain surrendered, Corner had implored the British Governor of Singapore to write to the invading Japanese commander and urge him to ensure the safety of Singapore's scientific and cultural wealth. His pleas succeeded. Corner hand-delivered the note. It was Japan's Emperor Hirohito—a marine biologist and amateur botanist—who gave the order that Corner and his two colleagues should remain at work, under the watch of two Japanese botanists. Together they protected the garden's valuable resources in the name of science, not nations. Corner went further. He gave secret support to those interned in the camps and smuggled to safety large collections of books from private libraries, and artefacts from the Raffles Museum.

Even back in 1945, Corner's efforts were recognised, but not publicly. A report sent that year to the British Colonial Office stated: 'The action of these officers in remaining at their scientific posts despite the adverse view of this which inevitably arose among those who were interned, has had results of the utmost value and scientific importance, and is to be highly commended.' No commendation was forthcoming. The facts only came to light in 2001, five years after Corner's death. Only then was he exonerated.

After the war, Corner worked for a while in Brazil before returning in 1948 to England, where he became a lecturer in botany and later professor of tropical botany, at Cambridge University. It was from there that Corner's work on *Ficus* could take off. He led expeditions to explore forests in Borneo and the Solomon Islands, where he found a fig species whose leaves were as long as he was tall. On Bougainville Island, east of New Guinea, Corner collected specimens from 40 species of *Ficus* in one small valley, where one tree in five was a fig. He more than doubled that tally on Borneo's tallest mountain, Mount Kinabalu, where he found 82 *Ficus* species.

In all, Corner had seen more than 300 fig species in nature. He combined his forest knowledge with studies of more than twenty thousand *Ficus* specimens in botanical repositories called herbaria, where he stared through microscopes for hours at leaves and twigs and figs, looking for tiny differences between species. He was thwarted in the 1950s, however, when he visited the herbarium at Utrecht University. Corner had wanted to look at the original *Ficus sumatrana* specimen the Dutch botanist Friedrich Miquel used when he named the species in 1867. When Corner opened the envelope that should have contained the preserved figs, he found only the stale remains of a cheese sandwich.

But Corner's bread and butter was botany, and it would take more than some missing figs to slow him down. By 1965, his knowledge of figs had coalesced into a landmark paper. It listed and explained how to distinguish 480 *Ficus* species that live in the arc stretching from India to Australia.

To achieve this, Corner cut down a virtual forest of more than 2,000 false *Ficus* species—names botanists had created for fig species that other botanists had already named. Corner's document would guide generations of later biologists, myself included. It is just part of his immense legacy. He was also a pioneer of conservation, who lobbied successfully to protect large areas of tropical rainforest. And his exquisite writing for non-scientific audiences engaged thousands of readers with nature. But Corner's brilliance had an evil twin. A fault line cut the rock of his character in two.

Even Corner's friends would say he was spiky, imperious, unforgiving. He could flip. Many suffered as a result. And when it came to work/life balance, Corner placed all of his weights on the work side of the scales. His family suffered and his marriage disintegrated. His children bore the burden of it all, and especially his son, John, whom Corner refused to see or speak to for decades. Yet it is thanks to this estranged son and his extraordinary book— *My Father in His Suitcase*—that much of the truth about this complex man is known.

Corner's personality spilled over into the story of *Ficus* when he fell out with botanist CJJG van Steenis, the editor of what Corner intended to be his masterpiece, the most comprehensive publication on fig species ever. The rift between the two men was so great that 35 years and Corner's last breath would pass before the text appeared, in 2005. This great tome, completed after Corner's death by Dutch botanist CC Berg, runs to 750 pages. It is a direct descendent of those sparse notes that Theophrastus, the father of botany, had written about fig trees nearly 2,400 years before.

It is fitting then that Theophrastus is recognised among the fig species whose names Corner retained. In 1868, a German botanist called Berthold Carl Seemann had named a fig species from the Solomon Islands after him: *Ficus theophrastoides*, a tree that grows just five metres tall and has such a slender trunk that you could encircle it with your hands. It is a far cry from the banyan, the species Theophrastus never saw but knew to be a fig.

Theophrastus had noted the thing that binds *Ficus* species together, their defining trait: the fig itself. But he never realised that the fig is in truth not a fruit. That fact, which helps explain the power of the fig trees, would elude science for hundreds of years. What Theophrastus didn't realise is that these magnificent trees have a biological secret.

It is a secret that solves the riddle of the *udumbara*, a species of Asian fig tree (*Ficus racemosa*) that Buddhist scriptures say flowers only once every 3,000 years. In fact, an adult *udumbara* flowers at least once a year, but like all fig species, it keeps its flowers hidden from human eyes. These flowers are for fig-wasps, tiny insects whose story is surely one of the most astounding in all of biology. It is one of the most important as well, for without these tiny creatures and their interactions with fig trees, the world would be very different.

SIX

Sex & Violence in the Hanging Gardens

On a moonlit night in southern Africa, a reproductive race is about to begin. The stakes are high but so are the risks. Most of the competitors will be dead or doomed by dawn. The starting line is a solitary fig tree whose gnarled form towers over a small stream. Figs hang in clumps from its branches like a plague of green boils. Tonight they erupt with life.

An insect emerges from a hole in one of the figs. She's so small you could swallow her and not notice. She's a fig-wasp with an urgent mission and her time is running out. All around her, thousands of her kind are crawling out of figs. Each one is a female with the same quest, and each faces immediate danger. Ants patrol the figs, and they show no mercy. Their huge jaws will crush and dismember any fig-wasp that delays her maiden flight.

Our fig-wasp avoids this fate with a flap of her wings that

lifts her clear of the carnage. She carries inside her body a precious cargo, hundreds of fertilised eggs that she can lay only in a fig on another tree. But she is fussy. The fig she seeks must be from the right species of *Ficus,* and it must be at the right stage of development. If it is ripe, she will be too late. If it is too small, the fig will not let her enter. The nearest fig that fits the bill could be tens of kilometres away.

The wasp does not have time on her side. With every minute that passes her energy stores deplete and can never rise again, for in her short adult life she never once eats. She has less than 48 hours to complete her mission, and although she has left the ants behind, the air brings fresh danger. Out of the dark night swoop bats, their mouths agape, their stomachs empty and expectant. The bats fly looping sorties through the clouds of dispersing wasps, condemning those they swallow to an early death. Our wasp escapes only when a gust of wind blows her high into the sky. She has eluded the predators. Now she must face the elements.

The fig-wasp is less than two millimetres long and her wings are thinner than a human hair. But relative to her body, they act as huge sails. With them, she rides the wild winds in search of a fig. She relinquishes control. Her fate is random now. Some of her cohort will be lucky and find their target within the hour. Many more will drop out of the sky, dead from exhaustion. The wind buffets her this way and that. All the while she awaits a signal from below, for the fig-wasp has allies in the trees she seeks. The trees need the wasps just as much as the wasps need their figs. Fortunately for both, fig trees are great chemists, and this makes them great communicators. At just the right time, they pump into the air a cocktail of chemicals that is unique to each species of *Ficus*. These compounds act in concert, like a choir of distinct voices that calls out 'welcome' in a language only certain kinds of wasps can understand.

There it is — a whiff of the perfume she seeks. As soon as she recognises it, she seizes control of her destiny and drops down out of the sky. She has found a patch of forest. Somewhere within it is a fig tree whose figs emit the signal scent. Away from the wind, she must now rely on her weak wings to carry her to the odour's source. The tree's figs are just right — smaller and harder than the one she departed. The fig-wasp has found her target but she has no time to rest. The final centimetre of her immense journey is among the hardest.

At the tip of the fig is a tiny hole. The fig-wasp squeezes her head into the hole and, with a resolute push from her slender legs, she forces herself forwards into darkness. The narrow tunnel in which she finds herself is tight. As she struggles forwards, its walls snap her antennae and wrench the wings from her back. It does not matter. This is a one-way journey and she will not need them again. She has come to give life but also to die, deep in the hollow heart of this special kind of fig.

The fig-wasp has other anatomical adaptations to help her reach her goal. Her head is shaped like a flattened wedge, ideal for forcing her way into the fig. Her jaws bear tooth-like ridges that dig into the tunnel walls. By opening and closing her mouth, the wasp ratchets herself forwards. At last she reaches the fig's hollow centre. She can complete her mission. And though the darkness blinds her, she knows exactly what she must do, in these, the last hours of her life. If her genes are to have a chance to survive, she must start to lay eggs, for the cavity at the centre of the fig will be both her tomb and her offspring's nursery.

Our wasp belongs to a species scientists call *Ceratosolen arabicus,* and her partner is the sycamore fig (*Ficus sycomorus*). This tree reaches up to 25 metres in height, with a dense crown of leaves that can spread twice as wide to form a canopy the sun's rays struggle to breach. The sycamore fig grows wild across a great swathe of Africa. Wherever it grows, it has become embedded in local cultures, often as a symbol of peace and unity—a place elders go to settle disputes. It is an ironic choice of icon, for figs are violent places. Within these 'fruit' you can find parasites that feed on living flesh and assassins that can survive only by killing babies. Figs are arenas of deadly gladiatorial battles and hasty incestuous sex. As biologist Bill Hamilton noted, in just one day as many as a million insects can die violent deaths inside the figs of a single tree.

Our fig-wasp's quest to reproduce does not end when she finds a fig where she can lay eggs. She has enemies ahead. She is deep inside the fig now. Flowers line its entire inner surface. They are packed together, their heads forming a carpet on which the wasp walks. As she does, she deposits pollen she has brought with her from the fig of her birth.

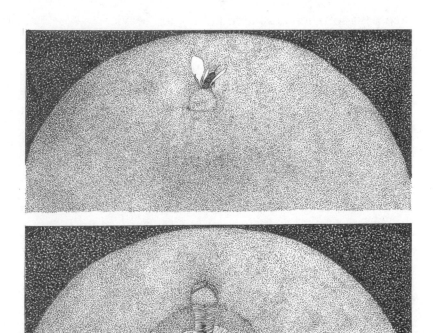

JOURNEY'S END:
A fig-wasp enters a fig through a tiny hole, then forces her way to the fig's
hollow, flower-lined centre

Each flower she pollinates can develop into a miniscule fruit with a single seed, an embryonic *Ficus sycomorus* that has the potential to grow into a giant tree. But not every flower shares this fate. Some of the fig's flowers will produce a new wasp instead of a seed. This is the price the fig tree pays for such a reliable pollination service.

To take her payment, the mother wasp gets down to the urgent business of laying eggs. One by one she penetrates the fig's female flowers with a flexible, needle-like structure at the end of her body. Through this hollow tube, she injects an egg into the part of the flower that would normally produce a seed. Each time she lays an egg, she also injects a drop of fluid. This induces the flower to develop a growth called a gall that will enclose and sustain her offspring. The larvae that hatch from her eggs will feed on the plant tissue in their galls until they are ready to metamorphose into adults.

The mother wasp must work fast. Her energy reserves are running low, and she has competition. Others of her kind have arrived and they too covet the limited supply of flowers. If she is fast, our wasp can lay more than 200 eggs. Finally, exhausted, she dies. Her final act will help ensure the fig species survives. And because of this, the tiny wasp will affect the fates of thousands of other species, all bound up in an intricate web of interactions that connects plants and fungi, microscopic mites and parasitic worms, birds and bats, monkeys and apes—and even you and me.

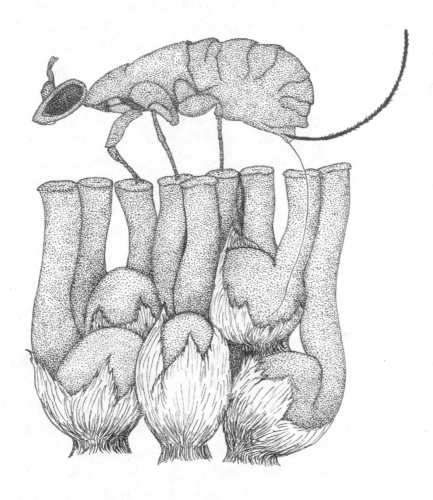

LIFE IN A DEAD-END:
Now wingless, a female fig-wasp pollinates fig flowers and lays eggs in some
of them through a tube at the end of her body

Fig-wasps have occupied enquiring minds since ancient times. Nearly 2,500 years ago, the Greek historian Herodotus wrote about 'gallflies' he found inside figs. A little later, in 350 BCE, his compatriot Aristotle described fig-wasps that emerged from figs and penetrated unripe ones. Another 50 years on, and Theophrastus was making notes on the insects he observed on figs. But it would then take eighteen centuries and the invention of magnifying glasses before anyone described fig flowers, which German botanist Valerius Cordus was the first to do, in 1544. That news did not spread. More than 200 years later, botanists still believed figs to be flowerless plants that reproduced with spores—as mosses and mushrooms do.

Only in the twentieth century did the quest to understand figs and their wasps really take off. A small but industrious band of biologists has spent decades examining the relationship in ever closer detail. The mathematical beauty they found in the way figs and their wasps interact matters to us because, while everything in nature is ultimately connected, the fig trees and their wasps appear to be disproportionately important cogs in the clockwork of life.

These biologists have shown that each of the 750 plus *Ficus* species depends on specific wasps to pollinate its flowers. For many *Ficus* species, just a single species of wasp performs this service. Less commonly, two or more wasp species do the job. This relationship liberates *Ficus* species from a constraint that limits other plants. It enables them to persist even at low densities yet maintain high genetic diversity because their partner wasps can carry pollen for as far as 160 kilometres—ten times further than the insect pollinators of any other plants. The relationship also has a critical side effect, ensuring a year-round supply of figs for hungry animals.

But the partnership between figs and their pollinators is under constant pressure from other tiny wasp species that are parasites of the relationship. Some sneak into a fig and lay eggs in its flowers but bring no pollen in return for the nursery the fig provides. These interlopers deprive the pollinators of places to lay their own eggs, whilst preventing fig seeds from developing in each flower they exploit.

Other wasps disturb the relationship between figs and their pollinators in more wicked ways. These parasitic wasps do not even need to enter a fig to cause trouble. They inject their eggs from outside—through delivery tubes that are several times longer than their bodies—and their offspring feed on the living flesh of the pollinator wasp larvae. Each *Ficus* species hosts up to 35 species of non-pollinating wasps. Overall, hundreds of wasps of various species of pollinators, seed-parasites and pollinator-parasites can develop in each fig. The outcomes of their interactions have impacts that ripple across entire ecosystems.

After a few weeks, the surviving offspring of the pollinator wasps are ready to emerge from their galls as adults. The males come out first. Their time on Earth is short and conducted almost entirely in darkness inside the fig of their birth. They have adapted to this lifestyle so much over millions of years that they are barely recognisable as members of the same species as the delicate females. A male fig-wasp's eyes are tiny or even absent. The males also lack wings. Another big difference between the sexes is in their mouthparts. The female never feeds as an adult, relying instead on energy she has stored after eating plant tissue as a larva. This will serve her well on the long journey ahead. It also means she has been able to evolve a slim head that makes it easier for her to enter a new fig.

The males, in contrast, need strong jaws. When they reach adulthood, they chew their way out of their galls then use their stout legs to dig through the dense thicket of fig flowers to find a gall with a female wasp inside. The males gnaw holes in those galls then curl their telescopic abdomens under their bodies to penetrate the galls and deliver sperm to the females trapped within. The males then move on, mating as quickly as they can and with as many females as possible, even if this includes their sisters.

The female wasps are soon ready to fly off in search of a fig in which to lay their fertilised eggs, but before they leave, they take receipt of the pollen they must deliver to their destination fig. In some fig-wasp species, this is a passive process. The fig's male flowers shed clouds of pollen grains that rain down and adhere to bodies of the wasps. But our *Ceratosolen arabicus* belongs in the other class of pollinators, the active ones. What these wasps have evolved to do is extraordinary.

After mating inside the *Ficus sycomorus* fig, the male *Ceratosolen arabicus* wasps use their massive jaws to chop down the pollen-bearing parts of the fig's male flowers. The female wasps then harvest the pollen with their forelegs, which are bedecked with stiff bristles like those on a broom. They sweep the pollen into cavities on the undersides of their chests called pollen pockets, which yawn open as they flex their bodies.

The female wasps are now laden with both eggs and pollen, but they are trapped inside their figs. To release them, the males, in their last act, do something otherwise unheard of in the insect world. They cooperate even when there is no direct benefit to themselves. These little males team up to chew a hole in the wall of the fig through which the females can escape. The males crawl out and die when they tumble from the fig or find themselves in the jaws of a predatory ant. In some fig-wasp species, the males appear to protect the departing females by sacrificing themselves to the ants. Off the females fly in search of the special scent that only their kind of fig emits.

The fig tree's pollen disperses into the night sky, bound to the bodies of its courier wasps. Thanks to these winged mediators the *Ficus sycomorus* tree may mix its genes with hundreds of other individuals. But this is only one side of the tree's reproductive effort. While some *Ficus* species have separate male and female trees, in others like *Ficus sycomorus,* each tree performs both sexual roles. Long before our tree's male pollen departed, its female ovules combined with the pollen its wasp partners brought from other trees. And so seeds formed alongside the new generation of fig-wasps, and soon they too are ready to disperse.

The tree's figs change in function, from incubators of wasps into attractants aimed at far bigger animals. They swell until they are three centimetres across and change in colour from buff-green to yellow or red. The plant withdraws its sticky latex from the figs and pumps in sugars instead. Birds, bats, monkeys and other animals come to feed on them. They will disperse the tiny *Ficus sycomorus* seeds that create a new generation of giant trees. Those trees will harbour future generations of fig-wasps and will feed future generations of seed-dispersing birds and mammals.

Without figs and their fig-wasps, many of these animals would starve. That's because most plant species produce their fruit at a specific time of year—often when many other species fruit too. This means fruit-eating animals experience periods of feast and famine as the amount of fruit in an area peaks in just a short period. But if all members of a *Ficus* species produced their figs at the same time, the short-lived female wasps that emerge from the figs would have no new immature figs in which to lay their eggs. It would mean no more pollination. This would doom both wasp and tree species to extinction.

Instead many *Ficus* species produce figs all year round, never all at the same time, and individual trees can produce two or more crops each year. Each day, the figs and their wasps introduce new beats to a rolling rhythm of fig production. It is one of nature's coolest tunes. It offers a lifeline to wild animals and so places *Ficus* species at the centre of vast ecological webs. The birds and mammals that eat figs will also disperse the seeds of many other plants whose fruit they eat. It is because of this that ecologists have described figs as keystone resources in tropical forests.

A keystone on a bridge or an archway locks all of the other stones into position. Remove it and the structure will come tumbling down. Remove keystone figs from a tropical rainforest, the analogy suggests, and this could trigger a cascade of local extinctions as birds, monkeys and fruit bats starve and are no longer around to disperse the seeds of thousands of other plant species.

In 1986, John Terborgh, then a biology professor at Princeton University, suggested that if figs disappeared from Peru's Amazon basin, the entire ecosystem could collapse. Later studies have identified a keystone role for figs in other forests, from Panama to South Africa to Malaysia and Indonesia. Biologist Daniel Kissling showed that across all of sub-Saharan Africa, the number of *Ficus* species in an area was the main factor affecting how many fruit-eating bird species lived there. Kissling concludes that figs are keystone resources on a continent-wide scale.

Right now, as you read these words, fresh dramas are playing out at fig trees across the tropics and subtropics, just as they have done every day for tens of millions of years. At some trees, fig-wasps are emerging from their figs and setting out on their bizarre and fatal journeys. At other trees, fig-wasps are arriving, bearing pollen and eggs. Without these ancient odysseys, the world would be utterly different.

For from the wings of tiny fig-wasps hang the fates of hundreds of bird and mammal species, and perhaps even entire rainforests.

SEVEN

Struggles for Existence

A mother-to-be is alone, hungry and helpless, jailed in a dark cell of her own design. She is a rhinoceros hornbill, the most striking of all Borneo's wildlife. This swan-sized bird wears a cloak of coal-black plumage from which bursts a figment of fire. It is the bird's casque, a curious horny ornament that curves skyward from the base of her huge bill and burns a flaming blend of red, orange and yellow. When the sun is behind it, her hollow casque can seem to glow like a hot ember. It sits atop a long bill that arcs down from its deep base to its sharp tip, fading from yellow to ivory-white as it goes. She is rainforest royalty. But for now her majesty must reign inside and unseen.

Days earlier, she forced herself into a dark cavity in the trunk of a tall rainforest tree. Her mate flew up to the hole and clung to its edge with his toes, fanning his white tail feathers against the tree for support. The pair then used their beaks like trowels to seal up the hole. For cement, they mixed a grim slurry of mud, regurgitated figs and fresh faeces. They ceased only when nothing but a small vertical slit remained unfilled. In the darkness, the female heard a telltale *whoosh . . . whoosh . . . whoosh*. It was air rushing through her mate's wing feathers with each beat of his departure.

When Tim Laman first heard that sound, in 1987, his life tilted in a new direction. The previous year he had been far away in the United States, a postgraduate student of neuroscience and animal behaviour at Harvard University. Laman was itching to escape lab work when he saw a curious poster in the biology building. The black-and-white illustration showed an orangutan in an urgent tussle with a spear-toting man. Laman didn't yet know the image came from a book by the cofounder of evolution Alfred Russel Wallace, his future hero. What caught Laman's eye next was the text: 'Wanted: Field Assistants for Rain Forest Research in Borneo. Contact Prof. Mark Leighton.'

Mark Leighton had spent several years studying the rainforests of Indonesian Borneo. He had shown that fig trees were important food sources for hornbills, orangutans and many other wildlife species. Once, he watched as a rhinoceros hornbill plucked, tossed, caught and swallowed 27 *Ficus binnendykii* figs in a minute. Now he wanted help at a research camp he had established deep inside Gunung Palung National Park, a vast area of rainforest. Leighton needed someone to manage the camp, carry out censuses of fruit-eating animals and track patterns of fig and other fruit availability. Laman got the one-year job and put his doctorate on hold.

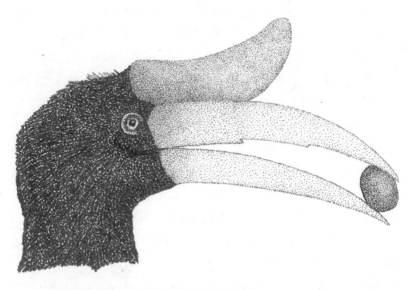

ANCIENT ALLIES:
Rhinoceros hornbills depend on figs to survive; in return for their food, they
sow *Ficus* seeds as they fly

The forest was pristine. The nearest people were several hours away, and the wildlife was abundant. It was on Laman's first day there that he heard that whooshing noise. He looked up and caught a glimpse of a massive black bird flying far above him at the roof of the forest. A flash of red told him it was a rhinoceros hornbill. Laman was an avid photographer but had no chance of taking a picture from the ground. From that day on it became his 'overwhelming obsession' to find a way to get up into the canopy with his camera.

Laman's year off became permanent. He fell in love with the forest, switched departments at Harvard and became a fig biologist. He attempted to answer a question that had vexed EJH Corner. In 1940, Corner had noted that strangler figs grew on only a small proportion of rainforest trees, yet in parks and fields they grew 'on every tree that is a roosting place for birds'. Laman wanted to know what challenges these stupendous plants must overcome to reach adulthood.

The national nark had 28 species of strangler figs. Laman focused on five of them and showed that each prefers host trees of a certain size and type, and has an ideal height at which to colonise its host. Some preferred to start out in life on big trees, high in the canopy. Others grew on smaller tree species and did best when they established in the gloomier sub-canopy. These differences limit competition among the *Ficus* species and mean more of them can coexist. But they also impose big limits on the strangler figs, by draining the pool of potential host trees for each species.

And as Laman would show, it is not even enough for a fig seed to land at the right height on the right kind of tree. Of the 134 strangler figs he studied, most had colonised their host in one of three places: the crotch between a branch and the host tree's trunk; a bulge on the trunk; or in a knothole that formed when a branch fell off. To learn what else limits these species, Laman had to emulate the birds and mammals that disperse their seeds. He focused on *Ficus stupenda*.

Unlike the strangler figs that can outlive their support trees and become free-standing, *Ficus stupenda* remains forever dependent on its host. Rather than encasing its host in cascading roots, this fig species tends to send a single root down the host's trunk and into the soil of the forest floor. As this pillar root thickens, it produces smaller side roots, which wrap around the support tree creating a bond between plant species that only death can break. As Laman found out, *Ficus stupenda* prefers to grow on the 60-metre-tall trees in a family called the dipterocarps. So these giants were the trees he had to climb.

Laman used a compound bow, fibreglass arrows and a reel of monofilament fishing line to send a climbing rope up and over big branches on such trees. Once his rope was secure and he had attached himself to it with a harness, he could pull himself up with a set of mechanical ascenders. Before he started, he donned thick gloves and went through his mental checklist. The tree was safe and his equipment well-maintained. The main danger now was human error.

Laman inch-wormed his way up to about 30 metres, the height he had found *Ficus stupenda* most often established itself and thrived. He took his time. He didn't want to aggravate the snakes, wasps or aggressive ants he was likely to meet. Once he fell foul of a bellicose bee that stung him right between the eyes as he hung from his rope. It was hot, sweaty and tiring work, but it was always worth it. From his vantage point, he could survey the roof of the rain-forest. At last, the rhinoceros hornbill was in reach of his camera's lens. Laman began developing the skills that would make him one of the world's foremost wildlife photographers.

In all Laman hauled himself up 45 big trees and planted nearly 7,000 *Ficus stupenda* seeds on their branches, crotches, knot-holes and trunks. He climbed up again 18 days later to check on his seeds, then again after seven months and once more after a year had passed. Laman had blessed those seeds with the right sites at the right height on the right trees, but their struggles had only just started. Eleven of the trees were already home to a species of ant for which Laman's seeds were the perfect takeaway meal. They carried many of the seeds off to their nests. Of the remainder that germinated where Laman had laid them, many then dried out and shrivelled and died. Others succumbed to insects that came to munch at their leaves. Only 1.3 per cent of the seeds survived a full year, and fewer than one in a thousand showed vigorous growth.

The seeds that fared best were those Laman planted on wood that was rotting. Soil and moss were the next best substrates. Leaf litter and bark were the worst. This all suggested that moisture, and not light, was the most vital factor. Laman confirmed this with a neat parallel experiment. He hoisted into the canopy ten lengths of plastic roof gutter and lashed them to his study trees. Each gutter contained moist soil and a pair of three-week-old *Ficus stupenda* plants, which Laman had grown from seed. His gutter figs, with their water-retaining soil, fared far better than the seedlings he grew on the host tree.

Laman confirmed that Alfred Russel Wallace had been right about the benefits strangler figs gain from germinating high in the forest canopy. Light helps them to grow faster. But what matters most is the presence of a substrate that can retain water, such as soil, rotting wood or moss. *Ficus stupenda* faces a challenge to ensure its seeds reach the rare knotholes and other crevices where moisture can accumulate. To deliver its seeds to these specific sites, a parent tree must employ couriers. It pays them with its figs.

Until now, its figs have been incubating a new generation of pollinator wasps. When those wasps have departed and her seeds are ready to disperse, the figs undergo a remarkable transformation. An influx of sugars sweetens their flesh. The sticky protective latex that had rendered them unpalatable retreats. A chemical blush turns the green figs first orange then red. She's letting the animal world know a feast is imminent.

While each fig species can rely on its pollinator wasps to be faithful partners, it engages with less trustworthy animals to disperse its seeds. And while *Ficus* species pay their pollinator wasps only after they have done their work, they must pay potential seed dispersers up front, so risk being exploited. A *Ficus stupenda* has no way of telling if it will attract paying diners, thieves that eat and run, or worse, killers with appetites for destruction. It's a lottery, but each *Ficus stupenda* invests in hundreds of thousands of tickets. If its pollinators have done their work well, each of its tens of thousands of figs will hold 160 or more seeds.

The first challenge these seeds face is to be eaten and survive the experience. It's a crazy way to start out in life. The fig flesh is there to entice animals to swallow the seeds. It helps that the seeds are tiny, just a couple of millimetres long. Even the smallest of fruit eaters can swallow them and not choke. But many of the animals that arrive are piratical. Green pigeons will gobble plenty of figs but the grit in their gizzards grinds each seed into dust. Parrots destroy fig seeds too, their horny tongues crushing them against their hard bills. More seeds fall prey to fig-eating squirrels. In each case, these animals are predators of *Ficus stupenda* seeds.

The seeds will survive a journey through the gut of most other fig-eating animals, but the smaller of these species — birds such as flowerpeckers and bulbuls — cannot open their gapes wide enough to swallow a *Ficus stupenda* fig. Instead they peck at the flesh, leaving most of the seeds untouched. These species are good dispersers of other fig species but aren't much help to *Ficus stupenda*. Nor are the animals that swallow fig seeds but live at lower levels of the forest and return there from the canopy soon after their feast. Any seeds dropped there will be too low. These creatures exploit the generosity of *Ficus stupenda* and provide little or nothing in return.

Another group of birds offers more. The barbets and the leafbirds, the fairy bluebird and the green broadbill are all fig specialists. They spend their days in and around strangler figs and swallow plenty of seeds, but they don't travel very far. What *Ficus stupenda* really needs are animals that eat a lot of its figs, move long distances, pass fig seeds intact and rarely travel below the forest canopy. Among the mammals, these seed dispersers include gibbons, orangutans and bats called flying foxes. Among birds, it is the hornbills, and especially the most impressive bird in Borneo, the rhinoceros hornbill. Here it comes . . .

Whoosh . . . whoosh . . . whoosh. The powerful wing beats announce the arrival of the sky king. The male rhinoceros hornbill in flight is a vision of sharp angles and exquisite curves. His long wing feathers splay like gloved fingers. His neck stretches out to support that huge beak and its fiery casque. Smaller birds scatter as he flaps up to the giant *Ficus stupenda*, spreading his white tail into a fan as he lands on a stout branch. He has flown three kilometres to find this tree, and he is hungry.

The oblong figs are on slender branches that would bend and snap if the three kilogram hornbill tried to walk on them. His beak, though, is long enough to reach them from a sturdier perch. Its serrated edges help him grip and manipulate each fig. With a quick backward jerk of his head, he flings the fig from bill-tip to gullet. He shuffles and

ALL HAIL THE SKY KING:
A rhinoceros hornbill arrives at a fig tree to feast

side-hops about until he has eaten all he can reach, then he flies on to another branch and starts again: pluck, toss, catch and swallow. But the figs are not only for him.

For 40 days now, after sating his appetite, he has returned from this and other fig trees to feed his partner in her hollow. She has laid two eggs. Soon there will be three hungry hornbills for him to feed. The male will bring lizards, mice and whatever fruit he can find. But it is ripe figs more than anything that he can count on. They are available year round, thanks to the partnership between fig trees and their tiny, short-lived pollinating wasps. The rhinoceros hornbills take full advantage. Studies have shown that figs can make up between 50 and 93 per cent of everything a female rhinoceros hornbill eats during her time at the nest. These birds eat figs from at least 24 *Ficus* species, fitting in other fare around this, their staple food.

Each day, the male bird delivers fresh rations of figs, poking his bill through the slit in the nest cover and coughing them into his mate's mouth. She is naked now, having shed all her feathers. If anything happens to him, she will become a skeleton entombed in the tree. For now at least, she is dry and safe from storms and snakes. From her secure cell, she may find solace in the periodic calls he makes to claim his piece of sky. It is a kind of nasal cough that can carry over hundreds of metres, a deep and reedy '*Hok! Hwok! Hok! Hok!*'

The male bird is helping more than his family. The figs stand to gain too. Laman calculated that half of all *Ficus stupenda* seeds fall beneath their parent tree's crown. For a species as picky as this about where its seeds germinate, that is just not good enough. It may seem improbable that a *Ficus stupenda* could ever reproduce, given its needs, but these are patient plants with ancient partnerships. The rhinoceros hornbill is an important ally that helps improve the odds of seeds reaching the right sites on the right kinds of host tree.

Adult hornbills can fly 10 kilometres in a day but must rest often on tall trees. For the hundreds or even thousands of fig seeds they excrete daily, these are perfect perches. A *Ficus stupenda* seed that falls from here has a better chance than most of reaching a moist crotch or knothole further down the tree. Once in a while, such a seed will survive long enough to germinate. Within days it will have sent forth roots to begin anchoring itself to its host. One root will begin its long journey down the tree's trunk to the forest floor. The fig's first pair of leaves will bask in the sun and power further growth. In time strong branches will spread wide, holding up a vast crown of leaves and fresh crops of ripe figs. They will provide food for future hornbills, birds that acquire a taste for figs before they even see the sky.

For after rhinoceros hornbill chicks hatch, they must remain in their tree hole. The adult male will continue to feed them and their mother for three months. Having regrown her feathers, she will then escape from her hollow, reseal the hole and together the parents will devote another three months to feeding figs to their chicks. It is rare for both chicks to survive, but the one that does will be strong.

If all goes well, this bird will live for nearly three decades. In the years ahead, it will feed from fig trees that grew from seeds its grandparents dispersed. In turn, it will spread seeds that grow into giant stranglers that feed its own grandchildren. At least, that is what has happened for millions of years. But the future of the partnership between *Ficus stupenda* and the rhinoceros hornbill is far from certain.

In the past century, loggers have carved up and cut down large areas of Borneo's forests. There are fewer tall trees for hornbills to nest in or for *Ficus stupenda* to colonise. The distances between fig trees and the hornbill seed dispersers they nourish has grown. The strain on their relationship is just one symptom of the disruptive effects of deforestation. When Tim Laman saw the threats habitat loss and hunting posed to Indonesia's forests and their wildlife, he felt compelled to do something about them. Writing scientific papers about a landscape that was so threatened frustrated him. He instead turned to another string in his bow. He reasoned that photos and stories could convince people to protect rainforests better than dry academic papers in science journals few people read. And he had the skills and the opportunities to take pictures nobody else could.

By 1992, Laman had climbed giant dipterocarp trees more than 500 times, but he wasn't only studying seedling strangler figs up there. He had found the trees rhinoceros hornbills would regularly visit to feed or display. Laman devised a way to build a hide from camouflage cloth, netting and leafy branches as he dangled from his rope high up a dipterocarp tree. Day after day, he would spend hours behind his hide while he waited for rhinoceros hornbills to feed on a nearby *Ficus*. His patience has yielded stunning portraits of these majestic birds.

By 1999, photography would be Laman's full-time job. He would go on to win numerous awards, but his big break had come in April 1997, when *National Geographic* published some of those hornbill photos in his first feature article: on Borneo's strangling figs. When I saw the magazine on sale in the Students Union at the University of Leeds, it felt like an omen. I had recently bought airline tickets—within a month I would be in Borneo, studying strangler figs, hoping to see a rhinoceros hornbill.

Three years had passed since I encountered the monk and his monkeys in Sri Lanka. I had completed my degree, worked for nine months in a bird garden, spent all my savings on a trip to the Amazon, then returned to university to study for a master's degree in biodiversity and conservation. My project supervisor was Steve Compton, the lecturer who had taught me the story of the fig-wasps. Steve arranged for me to visit fig biologist Rhett Harrison who was based in Lambir Hills National Park in Sarawak, one of two Malaysian states on Borneo. My mission was to study scores of wild *Ficus* species and the animals that eat their figs and disperse their seeds.

At just 70 square kilometres in area, Lambir Hills National Park secures only a small remnant of forest, but it is botanically rich. In just one 52-hectare patch, researchers have identified 1,178 tree species. That's more than in any other equal area of forest on the planet and more species than in all of the temperate forests of the northern hemisphere combined. In Britain there are just 36 native tree species. By contrast, in parts of Lambir Hills, if you enclosed an area of just 100 metres by 100 metres and identified every tree within it, you would find more than 600 tree species. Rhett had found more than 70 *Ficus* species in the forest. This diversity daunted me.

Few people alive know more than Rhett about Borneo's fig trees, and this book would not exist without all that he taught me about them. He was a great host. He's one of the most humane, humorous, and humble men I have met. Though he was still in his 20s, his hair was grey and this struck a contrast with his youthful appearance and propensity for cracking up in gales of laughter.

Rhett was born in Scotland but soon fell in love with the tropics. As an undergraduate zoology student, he led an expedition to a rainforest in Peru and ended up working just over the border in Bolivia for a spell to study mercury pollution. So he learned Spanish. His wanderlust led him next to Japan where he studied for a master's degree. So he learned Japanese. This degree and his later doctorate took him to Malaysian Borneo to study figs and their wasps. So he learned Malay and a lot of Iban too. While living in the national park, he had thrown himself into local life. As well as making his own rice wine, he attempted Iban weaving and performed the *ngajat*, a dramatic solo dance to the tune of gongs and drumbeats.

On my first night, over a dinner of barbecued fish, Rhett filled me in on the set-up there and his work in the forest.

As the sun set and the forest's frogs began to belch and bark, we drank his rice wine and gorged on durian, the stinking fruit that is a favourite food for orangutans. But as we chatted about his research, I started to feel out of my depth. It was complex talk that I did not fully understand. I started to wonder what I was doing in Borneo and whether I would be able to get a project out of my stay there. My doubts only grew the next morning when—after Rhett drove off to a wedding in a longhouse several hours away— his assistant Siba anak Aji arrived on his noisy motorbike and with him I entered the forest for the first time.

EIGHT

Goodbye to the Gardeners,
Hello to the Heat

'If you drop, you are dead,' said Siba anak Aji. I had met him just an hour earlier and already I liked his sense of humour. But he was right about the drop. A fall would provide plenty of opportunities to snap my neck. We were 30 metres high, dangling on a walkway that blazed an aerial trail through the rainforest canopy in Lambir Hills National Park in northern Borneo. The walkway was little more than a series of planks suspended in mid-air by a mesh of plastic coated cables anchored around big trees. With every step I took, the structure jolted, slid and creaked.

I tasted fear that first day. A safety harness tethered me to the walkway, but I did not trust it yet. Nor the insects. Little black bees hauled their bodies over my bare arms, thirsty for my sweat. Giant ants scuttled across my hands and boots. My skin crawled. What vanquished my nerves was the view. It was a vision of a distant past. Thick forests had dominated this landscape for a hundred million years. From the walkway, we could see the crowns of thousands of trees of hundreds of species. The tallest had burst through the canopy and reached 80 metres into the sky.

Colourful sunbirds and spiderhunters, barbets and flowerpeckers accompanied us as we traversed the 300-metre walkway. Squirrels crashed from tree to tree, their fur a blur of russet and cream. They sought what I sought—a pulse of life from the forest's beating heart. Siba found it first, a strangler fig whose branches bore thousands of orange figs. Within days they would be red and ripe. I would be shackled to the walkway, alone before dawn, waiting to discover what ate them.

In recent years, biologists had been saying big fig trees were the most important fruiting plants in the rainforest and should be conserved wherever possible. Siba had another reason these plants were special and must not be cut down. 'Spirits live there,' he told me. These two ways of thinking are linked. Taboos on felling figs have deep ecological roots—people who protect *Ficus* species benefit from the vast webs of nature these trees sustain. But as biologist Daniel Janzen wrote in 1979: 'A **fig** is not a *fig* is not a FIG', and Borneo's 160 *Ficus* species are more diverse than those anywhere else on Earth. I wanted to know what their variety meant, for both the *Ficus* species and for the animals that ate them.

Were all fig species important, or were some critical and others expendable? Such questions matter as we decide which aspects of wild nature to conserve and to which we wave goodbye. My short project grew into a doctorate, through which I would spend 18 months in the national park over the next three years. To find out what ate the figs of each of 34 *Ficus* species—a diverse mix of trees, climbers and giant stranglers—I watched them for hours on end, starting before dawn broke, then returning late afternoon and again at night.

Each time I stepped into that forest, I entered another world. It was hot and humid and full of mosquitoes. The forest's palette of greens and browns flooded my vision. Countless trees crowded in on me. Vines crept and corkscrewed their way skywards at every possible angle. Some were as thick as a thigh. Strange sounds tricked my ears. Strange shapes moved then vanished. There were musty scents whose sources I never found.

Most of the trees were just a few centimetres thick but were so numerous I could only take a couple of steps off a trail before hitting one. Others were giants, as broad as a small car. And none was as spookily beautiful as the first free-standing strangler fig I saw there. Its host tree had long since died and rotted away, and the strangler's roots now formed a scaffold with a hollow core. I stepped inside and looked up. Shafts of light shone down at me from far above. This *Ficus kerkhovenii* became my favourite landmark in the forest.

It soon grew clear that figs were ecological linchpins. In time I would record 49 of the park's bird species and 20

STRANGLER'S HEART:
When a strangler fig's host dies a hollow core remains

of its mammals eating figs. For each *Ficus* species, the size, colour and height of the figs determined which animals came to feed. While all figs were equal, some were more equal than others. I watched as rats gnawed on the tough brown figs that grew like huge warts at the base of *Ficus cereicarpa* trees. I stared through a night-vision scope at small *Ficus schwarzii* trees that thrust green chestnut-sized figs out of their trunks on short leafless stalks. Nothing but bats would dine on those scented figs. *Ficus punctata* meanwhile could count only on monkeys eating its red tennis-ball-sized figs, which were too big for most other animals to tackle. Another group of *Ficus* species, made up of slim trees and climbers with orange-red figs, attracted small birds and mammals that dwelt in the forest's lower storey.

The strangler figs were the pop-up restaurants of the rainforest. Their red, orange or purple figs attracted as many as 30 species of birds and mammals. The bigger the strangler's figs, the bigger the animals it attracted. These *Ficus* species operated on a boom-and-bust basis. They ripened as many as a million figs in just a few days and triggered a feeding frenzy that fell quiet as quickly as it began. Other *Ficus* species that grew as creepers or small trees drip-fed their seed dispersers, ripening just a small proportion of their figs each day for weeks or months on end. Some *Ficus* produced two or even three crops in a year. And thanks to the pollinator wasps, there were ripe figs every day somewhere in the forest. The daily pulses of fig production ensured that many animal species always had something to eat. Until, that is, the forest's heart stopped beating.

From January to April 1998, a severe drought parched all of northern Borneo. These months are normally among the region's wettest but only seven per cent of the average amount of rain fell. The drought wreaked havoc. The rain-fed rice fields of Siba's longhouse community needed daily downpours, but the soil there was now cracked and dry and thirsty. Across a large area of the national park, trees died at three times the pre-drought rate. For some species, the mortality rate shot up to between 12 and 30 times greater than normal.

Many of Lambir's fig trees died. The drought forced the survivors to make drastic changes to their biological business. To conserve water, they stopped producing figs. This deprived their fig-wasp partners of a place to lay eggs, and drove many of the wasp species locally extinct. As if the drought were not enough for the trees to contend with, fires broke out. They tore through the parched forest. The flames reduced part of my research area to a bleak landscape of charred, lifeless trees. All but one of the 43 *Ficus* plants in that area died. Lambir was not alone. Fires raged all across Borneo. Plantation developers on the Indonesian side of the island had started fires to clear the wild forest. The numbers of strangler figs, the *Ficus* species most important to wildlife, fell by 95 per cent in some areas of forest that burned there.

There is a fine line between resilience and vulnerability to such shocks. In Lambir, the drought receded as the rains returned. The *Ficus* trees that had survived began to produce figs again, but there were no wasps to pollinate their flowers. With no seeds to disperse, the figs didn't develop and dropped dead to the ground. Even six months later, most *Ficus* species were still without their pollinators. In time the wasps returned, aided by their prodigious ability to disperse over tens of kilometres. For many birds and mammals though, they came too late. The period without ripe figs had been deadly.

The drought was a relatively short shock, but the number and severity of these events has increased since 1970. If the trend continues, Borneo's forests and wildlife will face repeated tests. Another kind of change is already guaranteed. Our greenhouse gas emissions are set to increase the global average temperature by 1.5–4°C by the end of this century. In 2013, Nanthinee Jevanandam and colleagues at the National University of Singapore provided a chilling insight into what a warmer world could mean for fig-wasps.

In a laboratory, they exposed the pollinators of four *Ficus* species to rising temperatures and watched as the lifespan of all four fig-wasp species fell steadily. Jevanandam and colleagues studied wasp species from distinct branches of the fig-wasp family tree, so they think their results will be relevant to most other fig-wasps—many hundreds of species. They say global warming will reduce wasp lifespans and so limit the time the wasps have to find a fig of the right species and at the right stage of development in which to lay their eggs. If so, both the fig-wasps and the *Ficus* species they pollinate would suffer—with knock-on effects for the many bird and mammal species that rely on their bond.

Raise the temperature in a lab, and you can affect the chemistry within animals like these wasps. To increase the temperature globally makes all the world an experiment. Yet that is what is happening as a result of humanity burning fossil fuels and clearing forests. Will climate change spell the end of a relationship that has lasted 80 million years and forced thousands of species to dance to its rhythm? Fig-wasps may surprise us yet. After all, they have survived periods much warmer than today.

Steve Compton points out that the figs themselves may offer safe havens to heat-prone wasps. His doctoral student Areej al-Khalaf showed this with an experiment in Leeds, where Compton maintains a captive population of *Ficus montana* shrubs and their pollinator wasps in a huge greenhouse. Al-Khalaf waited until adult female wasps had entered figs to lay eggs in them, then she altered the temperature. Even a 10°C increase had no effect on the wasps' ability to reproduce and pollinate fig flowers.

Figs, it seems, provide their wasps with a relatively cool refuge. As these insects spend most of their lives inside figs, the real danger global warming poses will be to the short-lived adult females during their flights between figs. Steve told me he thinks fig-wasps could adapt to a rise in temperature, either in their physiology or their behaviour, by flying at a cooler time of day, for instance.

Right now, many *Ficus* species face more pressing challenges than rising temperatures. Unbroken forest once cloaked all of Borneo, but today only patches remain. Logging has heavily impacted 80 per cent of Sarawak's lowland forests. The habitat of fig species and the wild creatures they sustain has shrunk. Across the South China Sea in Peninsular Malaysia, biologist Andrew Johns found that the density of strangler figs fell by 74 per cent after supposedly low-impact 'selective' logging. Protected areas are meant to help, but in Lambir Hills, the tapestry I studied unravelled before my eyes. Science should be replicable, but if someone went there now the forest would tell them a different story, a drama with fewer actors.

By the time I arrived in 1997, the bigger fig-eating animals were already hard to find. I saw few hornbills despite spending thousands of hours in the forest, often watching crops of the figs these birds crave most. I saw only tiny gangs of green pigeons, a species that once flocked in their hundreds. I rarely saw monkeys. Gibbons, I heard only once and never saw. Where were all the barking deer and flying foxes, the sun bears and bearded pigs?

One answer became clear as I sat in the forest one day with Siba. On breaks he would offer me some rice wine from his flask and teach me the Iban and Malay names for the species pictured in my battered *Field Guide to the Mammals of Borneo*. This day the lesson was different. 'Boleh makan . . . Boleh . . . Boleh,' said Siba as he turned the pages. 'Can eat . . . Can . . . Can.' The list was long. The only animal categorically off the menu was the moonrat, a weird white-furred creature that stinks of ammonia. Everything else, said Siba, was fit for the pot.

Hunting was banned in Lambir Hills National Park and for Siba, and many other members of his community the park was a source of jobs, not meat. But for other people the forest was an unlocked larder. At night on the road that ran through the park, I often saw hunting parties from the nearby town of Miri. With powerful torches or spotlights mounted on trucks, they caught the eyes of deer and other creatures to shoot. I heard shotguns at night in the forest and found snares or camps that poachers had used.

The experience of two Malaysian scientists is telling. In 2004, Jayasilan Mohd-Azlan and Engkamat Lading set camera traps throughout the national park for a total of 1,127 camera-days. In that time they photographed only one bearded pig, once one of the commonest large mammals in Borneo. But the cameras caught four separate images of poachers, some toting shotguns. Three of the cameras were sabotaged—smashed or thrown away.

Two years earlier, Igor Debski and I had published a paper on the wildlife recorded in the national park, some 237 bird species and 64 mammals. While we added plenty of species to the list ourselves, many of those that people had seen before remained elusive. And since the 1980s, nobody had recorded three large and conspicuous species —a monkey called a banded langur, and two birds, the helmeted hornbill and great slaty woodpecker. This led us to conclude they might have gone locally extinct.

A decade later, in 2012, Rhett published research that painted a far worse picture. It described how surveys from 2003–2007 failed to find one in five of the park's bird and mammal species. The losses include half of the park's primate species and six out of seven hornbill species. If Lambir loses these animals, it will lose much more besides, for they are supreme seed dispersers not only of figs but also of many more species. Across Asia, hornbills disperse the seeds of more than 500 plant species. But it is the *Ficus* species they disperse that affect the most other species, most immediately. Between them, Borneo's eight hornbill species disperse the seeds of at least 45 different *Ficus* species, whose figs feed more than a hundred species of birds and mammals, which in turn disperse the seeds of thousands more plant species.

Early in 2012, two weeks before I heard about Rhett's new paper, I bumped into someone who used to work for the Sarawak National Parks and Wildlife Office. I told him 10 years had passed since I had been in Sarawak, and I asked him how the state's hornbills were doing. 'Who cares?' he said with a smile and shook his head. He cared of course, and he knew I did too. He meant it was already too late to care because no one with any power did.

In under 20 years, hunting has stripped Lambir Hills National Park of its bigger animals. When I worked there in the late 1990s, I could still find 25 to 30 species of birds and mammals feeding on the figs of a single *Ficus* tree over the course of three days. A decade later Rhett reported that the same *Ficus* species could only attract half as many animal species. Many of their figs fall uneaten. Their seeds are not spread. It's a pattern repeated across the tropics and it has a name: 'empty forest syndrome'.

It doesn't need to be this way. Rhett points out that Lambir's missing species seem to be doing okay in Sungai Wain, a similarly sized but far more degraded forest in the Indonesian part of Borneo where local hunting has been regulated. But in Lambir, the decline in numbers of large fruit-eating animals—the gibbons, monkeys, flying foxes and hornbills—spells trouble for the species whose seeds they disperse. And that includes the fig trees upon which so much else depends.

The irony is that the fig trees helped sow the seeds of this situation. Indeed, we can trace its origin back millions of years to a time when figs nourished our pre-human ancestors. These trees shaped the world in which we evolved and then fed our bellies and fuelled our imaginations for thousands of years through important episodes of human history. With the rise of humanity and our hands, hunger and hubris, it was only a matter of time before we took control. Our success has changed the rules of the game.

NINE

From Dependence to Domination

Dear Reader: You and I are related, both in blood and through figs. We share ancestors that survived and thrived because they dwelt among *Ficus* trees and ate often from their ripe crops. Figs helped make us. If we had a time machine, I could show you. The story begins 80 million years ago, when giant dinosaurs still roamed the world. The alliance between fig trees and their wasps had taken root. Our ancestors were small, furry creatures. Their prospects changed 66 million years ago, when the universe propelled an asteroid into the planet. It blasted a new future into being.

The asteroid was at least 10 kilometres wide and travelling 20 kilometres a second when it smashed into the Earth at the edge of what is now Mexico's Yucatan Peninsula. Shock-waves ripped around the world, sparking volcanoes and earthquakes and tsunamis that rose hundreds of metres into the air before crashing across coastlines with devastating force. A pall of dust and ash lingered in the atmosphere for years, blocking out sunlight and cooling the planet. Three-quarters of the planet's animal and plant species went extinct. The giant dinosaurs could not cope with the change. They fell like stones and exist now only as fossils.

For the fig trees, though, it was just the start of a new chapter in an already long story. They and their partner wasps survived the apocalypse. So did the mammals. They diversified into new forms and thousands of species. Figs would feed many of the newcomers, from rats and bats to antelopes and elephants. They became particularly important to the primates: the group whose modern members include monkeys, apes, and you and me.

Figs matter so much and to so many primates today that it seems likely they have been feeding our family for tens of millions of years. Our close relatives, the gorillas, chimpanzees and orangutans, all love a meal of figs. So do lemurs and gibbons and dozens of monkey species in Africa, Asia and South America. As our ancestors scampered along branches, then grew bigger and brainier and ultimately descended to walk upright, figs were rarely beyond their reach.

Evidence that figs sustained the ancestors from which humans evolved has emerged from a site called Aramis in the Afar desert, a bleak landscape in Ethiopia's Middle Awash region. It's an unforgiving place: hot, dry and largely barren. Any urbanite stranded there today would struggle to survive. But 4.4 million years ago this sandscape was a grassy woodland, green and moist and full of life.

The raucous calls of parrots and peacocks rang out from the woods each day. Porcupines and spiral-horned antelopes foraged in the undergrowth. Monkeys rollicked around in the treetops high above. Fossil evidence shows that fig trees grew in that woodland. Among the species they likely fed was *Ardipithecus ramidus*, a 1.2-metre-tall primate that may be the ancestor of us all.

Ardipithecus ramidus was not an ape, but a member of the human side of the family tree. It had a small head, long arms and very long fingers. It could walk upright better than a chimp and climb trees better than any human. Its big toes splayed out to the side of its feet, enabling it to grasp branches as it clambered about on all fours when up trees.

This species was unknown until 1992, when palaeontologists Tim White, Berhane Asfaw and colleagues began to crawl across Aramis, their eyes alert for tiny fragments of bone. First they found a single tooth and an arm bone. These finds were distinct enough from anything else known for the researchers to declare a new species but not enough for them to conclude much about how *Ardipithecus ramidus* lived. In 1994, though, the team struck fossil gold. They had unearthed the first of more than a hundred bones from a single *Ardipithecus ramidus*, a female they would call Ardi.

The fragile, off-white bone fragments were so soft they would crumble when touched. The researchers used dental picks, bamboo strips and even porcupine quills to expose the bones, then used plaster and aluminium foil to extract them from the sediment. It would take the team 15 years to clean, assemble, measure and analyse their finds. Not until October, 2009, could they publish their detailed description of Ardi and the landscape she inhabited. It was one of the greatest fossil discoveries of all time, a window into our evolutionary past. The researchers say that, if *Ardipithecus ramidus* is not our direct ancestor, she must have been closely related to it and would have been similar in appearance and adaptation.

The shape, size and structure of Ardi's teeth suggest she ate fruit, leaves and small mammals. On some days, she and her family would have woken with rocks of hunger rolling in their bellies. The fig trees Ardi lived among would have offered a lifeline. Just how important figs were to our ancestors that long ago we can only guess, but a sidelong glance at another relative may help. Ardi's brain was about the size of a chimpanzee's. While chimps are very different animals than Ardi was, they also live among trees and divide their time between the forest floor and the branches above. Chimps, then, may give us insights into how our pre-human ancestors interacted with fig trees.

Chimpanzees love figs. They pluck them from at least 30 *Ficus* species and eat them whenever they can. Richard Wrangham and colleagues at Harvard University got a sense of how important figs are to these animals during research in a forest in Uganda. Wrangham's team collected nearly two thousand piles of chimp faeces over a four-year period. For most of that time, from every single sample, fig seeds shone like little nuggets of gold.

Figs are a chimp's perfect source of energy. But in their

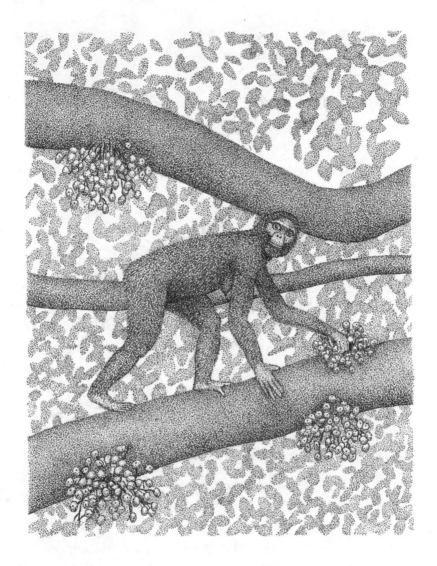

FIGS IN THE FAMILY:
High in an African fig tree more than four million years ago, an *Ardipithecus ramidus* reaches for a fig

daily quest to find ripe figs, chimps face challenges that would have dogged our pre-human ancestors too. Big fig trees are rare, scattered throughout forests and other landscapes at low density. Ripe crops appear randomly in time and space. The terrain is tough. Nor are chimps alone in their love of figs. Monkeys, hornbills and dozens of other animals all desire them too, just as they did in Ardi's time. Competition can be intense.

Chimpanzees overcome these challenges with brainpower. Biologist Emmanuelle Normand has shown that chimps are smart enough to remember where various fruit trees stand in a patch of forest. This means they can efficiently patrol their territory and never miss a ripe crop. In research published in 2014, Karline Janmaat and colleagues showed that chimps even plan where to sleep at night so they can beat other foragers to a breakfast of ripe figs. They set off earlier the more distant the figs are. This ability to weigh up information about when and where figs are available helps chimps beat their competitors to food supplies that are here today, gone tomorrow. It means they can secure a steady supply of the high-energy figs they need to fuel their large brains.

Research published in 2016 by Nathaniel Dominy and colleagues shows how chimps gain another important advantage over their competitors. Unlike other animals, chimps can use their fingers and thumbs to squeeze figs before deciding whether to eat them. Even monkeys cannot do this. The chimps are assessing softness which is a good proxy for how much sugar the figs contain, and therefore how much energy the chimp can gain from eating them. It means chimps don't waste time and effort biting into unripe figs. Ardi would have been able to do this even more precisely than a chimp. Dominy says dependence on figs may have played a key role in the evolution of manual dexterity in our pre-human ancestors, a trait that reached its pinnacle in the fully opposable thumb that sets our species apart from other primates and enables us to make and use tools.

Chimps may even use fig trees as pharmacies, according to Shelly Masi of the National Museum of Natural History in Paris. Working in a forest in Uganda, Masi and colleagues observed wild chimpanzees eating unusual foods, including the leaves and bark of five *Ficus* species. Although these foods were common, the chimps ate them only rarely. Masi's team thinks that this could be because the chimps ate these foods for their medical benefits, rather than for nutrition. The fig leaves and bark contain compounds that have shown effects in laboratory tests against bacteria, the malaria parasite, tumours and parasitic worms. Masi's study raises the tantalising possibility that our pre-human ancestors also visited their favourite fig trees for drugs as well as food.

I wonder what Ardi would have thought about those giant fig trees whose bark was smooth to her touch, whose soft nourishing figs she could so often count on finding. We don't know how important figs were to her kind, but I cannot imagine she didn't relish the sight of them. A creature like that was the ancestor of us all. She could hunger just like we do.

After Ardi died, more than four million years would pass before the first anatomically modern humans took their first steps. Throughout that time figs would have continued to feed our ancestral line, even as our branch of the family tree bent down to the ground. Our ancestors walked away from a life in the trees, but the trees of life, those beneficent figs, had taken root in their consciousness.

When the first people spread across Africa, they found figs almost everywhere they went—more than a hundred *Ficus* species inhabit the continent's forests, savannahs, swamps and drylands. The figs appeared year-round, a bounty they could count on. But fig trees also had other gifts. They provided medicines and shade and good hunting. Just as today hunters in forests across the tropics will stake out *Ficus* trees, the early humans would have realised these trees are meat magnets. A canny hunter with a rock or a spear could have taken out a monkey or a large bird for a double win, a mixed meal of figs and flesh.

In serving several basic human needs, fig trees began to cement themselves into culture. Over time and across the entire African continent, this would translate into myths, rituals and taboos. New links formed when people stepped out of Africa for the first time. As the wanderers reached the Mediterranean and the Middle East, then Asia and Australasia, and even in time the Americas, they found fig trees waiting with gifts to share. They created new stories in which these trees starred.

A turning point came around 12,000 years ago as people started to do something they had not done before. They stopped moving and settled the land. One of the first places they did this was in the lower Jordan Valley. Fig trees helped tether people there, says Mordechai Kislev of Israel's Bar-Ilan University. In 2006, Kislev and colleagues reported that they had discovered the part-fossilised remains of figs people had stored in a building around 11,400 years ago, alongside wild barley, wild oats and acorns. The figs were seedless. Kislev's team concluded they were from a sterile form of *Ficus* trees that could only increase in number if people planted branches in the soil. Kislev argues that people did this with intent, to create a local food supply, thousands of years before anyone in the region domesticated wheat or millet. He says they subsisted by foraging for wild seeds, nuts and meat, and harvesting figs from the trees they had cultivated.

It was a controversial claim. Other biologists have argued there is not enough evidence that the figs were indeed sterile. Even if Kislev is wrong, figs were still among the first plants that people domesticated. Other researchers have found evidence of fig cultivation dating back to the early Bronze Age, 5,000 years ago. It was only a matter of time before farmed figs were feeding the great civilisations of the Sumerians, Egyptians, Minoans, Greeks, Babylonians and Romans.

Humankind was making a critical switch. We had long danced to nature's rhythm, now we tried to set the beat. The twin forces of agriculture and urbanisation would alter us and our planet for ever. Back in Africa, fig trees would play key roles in the foundation of ancient Egypt. They provided timber for construction and boat-building. Their figs fed slave and pharaoh alike.

The species that sustained ancient Egypt was the sycamore fig (*Ficus sycomorus*). This huge tree can grow even at the edge of the desert, far from the Nile, thanks to extensive roots that seek out water deep underground. Yet a death sentence hung over these trees. In Egypt, the species was doomed. Its pollinator wasp was nowhere to be found. Each year the trees pushed out expectant figs, whose hundreds of tiny flowers never felt the feet of a pollen-bearing wasp. Without its wasp, a *Ficus* plant produces figs in vain. Without pollen, there can be no seeds, no need for dispersers. So the figs stay small, hard and green then drop to the ground where they rot. More than 5,000 years ago though, someone made a discovery that would subvert nature and change history. They had found a way to trick the figs with just a flick of the wrist.

Pierce an unripe *Ficus sycomorus* fig with a sharp blade, and the fig seems to think it has been pollinated and has seeds to disperse. In just a few days, it will swell and grow sweet. This ancient innovation gave Egypt's dead-end figs a new lease of life. Farmers could grow new trees by planting branches in the soil. Some farmers trained baboons to harvest the ripe figs, pre-dating EJH Corner's botanical monkeys by thousands of years. The trees would produce two, three or even more crops of figs a year. After being tricked into ripening, the figs could be eaten fresh or dried for later use. They provided a year-round food supply.

HEAVEN'S GATE:
The Egyptian goddess Hathor emerges from her sacred fig tree to meet souls
bound for the afterlife

The sycamore fig tree came to embody various goddesses, including Nut, Isis and Hathor. Each has been called
'Lady of the Sycamore', but it was with Hathor that the
connection was strongest. She and the mythic fig tree she
inhabited were milestones on the journey to heaven. Pharaohs were buried in coffins of *Ficus sycomorus* wood to
speed their return to the womb of this mother tree goddess.
They took to their graves figs from the same species to
sustain them on their journey into the afterlife. The pharaohs believed that after death their soul would encounter
Hathor in a fig tree that grew on the eastern horizon.
Beyond it was the underworld. Hieroglyphs inscribed
inside pyramids at Saqqara in about 2400 BCE describe how
Hathor would emerge from her tree to offer the dead pharaoh's soul figs, bread and water. By accepting these gifts
his soul would become her guest and feed for eternity on
the figs of paradise.

By the time of the Middle Kingdom (2040–1640 BCE),
this route into heaven was open not only to the pharaohs.
Egyptian commoners also believed that if they lived cor-
rectly, their souls could join Hathor in her fig tree. Beyond
the tree was a hellish swamp populated by dangerous
snakes and crocodiles. Fearsome giant baboons trawled the
murky waters with nets in search of food, including lost
souls. Every day the rising sun reminded the Egyptians
that the underworld lay beyond the fig tree. For it was
Hathor who made the day possible by giving birth each
morning to Horus, the sun-child who then rose to the fig

tree's crown before transforming into the sun god Ra and entering the sky.

For all that *Ficus sycomorus* meant though, it was a tastier cousin that would become the queen of figs. While the Egyptian figs were powering the arms that built the pyramids, farmers in what is now Greece were mastering *Ficus carica*, a species now grown for its energy-packed figs in more than 70 countries. Those figs are more than just a sweet treat. They are the original superfood. They are rich in vitamins and minerals such as calcium, iron and potassium. They also have more fibre than any other commonly grown fruit.

Around 2,500 years ago, the Greek figs had become so highly prized that the Athenian statesman Solon made it illegal to export them. Bans of course create criminals. In doing so, this ban also changed the Greek language and created a word we use in English today: sycophant. The word originally referred to the people who showed that figs were ripe and ready to eat. It later came to describe people who snitched to the authorities about anyone exporting figs illegally.

It wasn't long before people started to see figs as more than mere food. They noted that people who ate them fared well. Roman author Pliny the Elder wrote that: 'Figs are restorative . . . the best food that can be taken by those who are brought low by long sickness. . . . They increase the strength of young people, preserve the elderly in better health and make them look younger with fewer wrinkles.' The Prophet Muhammad later declared figs to have been heaven-sent. According to Islam's Hadith literature, he distributed figs to his followers and told them: 'Eat it as it cures various diseases.'

Meanwhile across Africa, Asia, the Pacific and South America, people made medicines from their own wild fig species. In India, there is a written record. Nearly 2,000 years ago, scribes there began to document the Ayurveda system of traditional medicine. Its remedies use extracts from several fig species to treat diarrhoea and constipation, mumps and jaundice, boils and haemorrhoids, diabetes and dysentery, open wounds and tooth decay. Today in Nepal, people use the leaves, bark and roots of just one fig species, *Ficus benghalensis*, to treat more than twenty disorders.

Among indigenous peoples who live in the forests of Guyana, French Guiana, Colombia and Brazil, there are more remedies. Some use the sticky latex that flows from the cut stem of a fig tree to treat swelling, cuts, sprains, fractures, abscesses and worm infections. Others use powdered ash of burnt *Ficus* stems to treat children with diarrhoea. Many of these traditional medicines are still in use today. Meanwhile, every week scientists publish new research assessing figs as potential sources of drugs for cancer and other diseases. It's a fusion of modern science with knowledge so ancient that it may well date back to before we descended from the trees.

For millions of years, then, our ancestors gained much from fig trees. They took from them food and shelter, medicines and materials—gifts that must on some days have seemed divine. Add to this the awesome appearance of many fig trees, their power to crush and conquer other trees, and their seemingly endless fertility. It's not hard to imagine why people would protect and even worship these trees — nor why so many cultures planted them in their soil and their stories.

Fig trees often also provide less direct benefits, as centrepieces of intricate natural webs. They support the seed dispersers of other plants that provide us with fruit, medicine and timber, and plants whose flowers feed the bees that pollinate our food crops. Nature would reward any community that protected fig trees, and many chose to thank their gods for those blessings. Others went as far as saying that these gods, or the ghosts of their ancestors, inhabited the trees. Often there is a clear ecological basis to such thinking.

In the Congo basin, Bakongo people founding a new village traditionally planted *Ficus thonningii* fig trees to assess whether a site had adequate water. If the fig tree thrived, this meant the water spirits were content and that other trees which supplied fruit or shade could follow. On the island of Pohnpei in Micronesia, the local name for a large *Ficus prolixa* fig tree means 'that which holds the land', a reference to the tree's deep roots, which limit erosion. These trees are also known as the home of spirits. In Guatemala, Maya communities believe that felling a *Ficus cotinifolia* will cause the rain to stop, as the tree is the source of water. Stories there tell how angels and ancestral spirits rested on a giant fig tree and woe befell anyone who tried to harm it. One man who tried to cut the tree had a stroke and was part-paralysed for life. Another tried to climb the tree and was turned into a monkey.

The traditions are dying. But go to the highlands of central Madagascar and you'll find the beliefs are still strong. Taboos are intact. The Betsileo people there are farmers whose villages dot a landscape of hills and valleys. In many a field, a giant fig tree stands tall. Some of these are trees the Betsileo left standing when they cleared the surrounding vegetation. Others, they planted in religious ceremonies. When biologist Emily Martin and colleagues went to study the value of these trees, they found that they enhanced the diversity of birds. Without the fig trees, they said, forest birds would not enter open farmland. The trees serve as stepping stones across which genes—not only of the birds but also of the seeds they disperse—can travel between distant patches of forest.

These isolated fig trees provide the Betsileo with shade, fruit, medicines and fibres, which people weave into baskets. The hungry birds and bats the fig trees lure are sure in turn to attract Betsileo hunters. The fig trees are clearly worth more alive than dead and so the Betsileo spare them. Over many generations, they have woven these figs trees into their rituals and taboos. Some fig trees represent ancestors. Others serve as markers of land boundaries, memorials to ancient villages or sacred posts for rituals. Anyone who chooses to cut one down risks misfortune. In the case of *Ficus tiliifolia*, this could mean blindness in a family member or a new mother failing to lactate.

'Don't cut down fig trees'—it's a message people across

the planet have repeated over millennia. This gem of wisdom pre-dates the dawns of agriculture and civilisation. From before the first humans left Africa, people would have known that a fig tree alive is a lifeline. In return for the gifts the fig trees offered, our ancestors provided protection and—for many of the *Ficus* species—a fine seed dispersal service too. Over millions of years, our family was a partner to the *Ficus* species. Today we destroy them with casual indifference. It's a reversal epitomised by the fortunes of fig trees in Kenya. These trees have played many curious roles there, from wartime lookout post to clandestine post office, from conduit of divine power to symbol of society. In this particular story, they star alongside a queen and a seer, a Nobel Prize winner and the most wanted man in the British Empire.

TEN

The War of the Trees

In the mid-1940s, a little girl called Wangari played beneath a giant fig tree near her home in the highlands of Kenya. 'That is a tree of God,' her mother told her. 'We don't cut it. We don't burn it. We don't use it. They live for as long as they can, and they fall on their own when they are too old.' At the time Wangari did not know why the fig tree was so special. She just knew the stream that flowed beneath it was a good place to hunt for tadpoles.

Six decades later, after battling corrupt politicians, enduring police brutality and winning the Nobel Peace Prize, Wangari Maathai would speak often of that giant *Ficus*. In the intervening years, she had learned why her ancestors had let fig trees stand tall. She had seen too what happens when these trees fall. And this had helped bring forth from her mind a 'little big idea'.

The fig tree was a *Ficus natalensis*, a species Maathai's Kikuyu people called 'mugumo'. These elephant grey trees are wonderful things. They can tower to more than 20 metres in height with an evergreen crown of leaves that reaches 30 metres across. As they grow their branches drop aerial roots which can smother other trees. Their thick trunks have supplied a local metaphor for the ineffectual: 'Like cutting a mugumo with a razor blade'.

A mugumo's small round figs ripen a warm orange tone, luring in hungry monkeys, fruit bats and birds including tinkerbirds and turacos. Elsewhere in Africa, *Ficus natalensis* figs feed our close relatives, the chimpanzees and gorillas. It's one of the species whose figs likely fed some of the first humans too. For Maathai's mother, though, a mugumo meant faith not food.

In the traditions of her Kikuyu culture, both the mugumo and another fig species, the mukuyu (*Ficus sycomorus*) could become sacred places. Her people are, in a literal sense, children of the fig tree. They trace their origin back to a founding couple, a woman called Mūmbi and a man called Gīkūyū, whose name means 'giant *Ficus sycomorus* tree'. Their god Ngai had told Gīkūyū to settle in an area rich in fig trees close to Mount Kenya, Ngai's abode on Earth. Ngai said Gīkūyū could commune with him in times of crisis by making offerings at the fig trees.

The couple soon started to produce children—all girls, one after another. By the time their ninth daughter had reached marrying age there was still no sign of a man to wed any of the sisters. So Gīkūyū and Mūmbi went beneath a fig tree, sacrificed a lamb and prayed to their god. When they returned to the tree the next morning, they found nine men who, upon marrying the first of the couple's daughters, gave rise to the first nine of the ten Kikuyu clans.

Over centuries, the wild fig trees insinuated themselves into the culture, religion and identity of the Kikuyu people. The mugumo trees provided fodder for livestock and medicines for people. With their milky latex and abundant figs, they were symbols of female fertility. 'May you be blessed by Ngai,' goes one Kikuyu saying. 'May you be as fruitful as the mugumo tree.'

With their massive size and power to dominate landscapes and other life forms, the mugumo trees became the people's prime symbol. Once consecrated as a holy tree, a mugumo was protected from felling. It became a site of sacrifices and ceremonies, from circumcisions to transitions of power and prayers for rain. The sacred tree was the meeting place that gathered people from their dispersed homesteads. For all of these reasons, the Kikuyu protected the wild fig trees. Everything changed when the Europeans arrived.

By the mid-1800s, missionaries were active in Kenya. Many of the Kikuyu converted to Christianity. Churches replaced mugumo trees as places of prayer, sometimes springing up alongside the sacred trees. One group of missionaries sent the wood of a fallen mugumo tree all the way to Scotland, where it was carved into a cross and returned for display in their church. Many of the converts retained some of their traditional beliefs, blending them into their new faith.

While the missionaries thought they were liberating local people, other Europeans came with brazen intent to exploit them. In 1884–5, when the European powers carved up Africa, Britain's slices included Kenya, which it claimed as a protectorate and made a colony ten years later. Over the decades preceding Maathai's birth in 1940, Kenya's colonial government had seized vast areas of land and allocated it to white settlers. Plantations replaced forests. Shabby little tea bushes usurped mighty ancient fig trees. The colonial administration forced the people it evicted into overcrowded reserves and enacted policies of forced labour, conscription and taxation that further alienated and impoverished the African majority.

By 1952, when Wangari Maathai was 12 years old, an armed uprising against colonial rule had begun. Most of the rebels were poor and landless Kikuyu. They named themselves the Kenya Land and Freedom Army but became known as the Mau Mau. The violent episode that was unfolding would set the stage for Kenyan independence. Historian John Lonsdale has called it a 'symbolic war of the trees'. For the Kikuyu especially, control over ancestral forests was at stake. Fig trees played many curious roles in the conflict. And few among the Mau Mau took their fig trees more seriously than their commander-in-chief, Dedan Kimathi.

Kimathi dubbed himself Field Marshal, and later Prime Minister of the Southern Hemisphere and Knight Commander of the East African Empire. In reality, he had worked at several dairies, a timber company and at a Shell oil depot. In 1940, he had enlisted in the British Army but was kicked out for violent conduct. Now though he could call himself what he wanted. He commanded fighters in the Aberdare and Nyandarua forests with magnetic oratory and threats of violence. They did his bidding with unflinching loyalty.

Before long Kimathi had a price on his head. The bounty was £500, a small fortune. Yet, even as he fled through Kenya's forested highlands from soldiers with orders to shoot to kill, Kimathi still took time to connect with the fig trees. One of his favourites had a thick trunk and massive weighty branches that hung down almost to the ground. Taking refuge from the moonlight beneath this tree, Kimathi raised his arms into the air and pressed his dread-locked head against its cool bark. He smeared an offering of sticky honey at the tree's base and prayed to his god Ngai to come and save him. His prayers over, he slipped away into the forest and the night, knowing that only death or victory lay ahead.

Nearby, and on the opposite side of the conflict, lived Eric Sherbrooke Walker, a retired British Army Major. Walker also knew a thing or two about being on the run, and he too had a soft spot for fig trees. He had served in both World Wars, and when captured by the German forces in the first, he had tried to escape 36 times. After World War I, Walker became a smuggler, trafficking rum from Canada to the United States of America during Prohibition. This funded his move to Kenya, where a giant fig tree would soon seize his imagination.

In 1932, Walker built a two-room tree-house in a huge mugumo tree in what is now Kenya's Aberdare National Park. It was an ideal spot for viewing the elephants, hippos

CROWNING GLORY:
Treetops Hotel, Kenya in 1952, when a princess climbed the fig tree and
returned to Earth the next day a queen

and other wildlife that visited a nearby watering hole. In time Walker developed this modest shelter into a luxury lodge called Treetops Hotel. On the night of February 5th, 1952, the life of its most famous guest—a 25-year-old English woman called Elizabeth—was transformed.

Back at home in England, the young woman's father, King George VI, had died in the night. Before she had descended from the tree she had ascended to the throne as Queen Elizabeth II. Along with the crown, she inherited the vast British Empire, whose subjects ached for independence. Nowhere was this truer than in Kenya.

Within months, conflict gripped the colony. The uprising had taken root. The rebels fought a guerrilla war, hiding out in the forested highlands, including the Aberdare range where Walker's hotel perched amid the fig leaves. Britain took to the air and bombed the forests to flush out the fighters. More than six million bombs fell from British planes in less than two years. As the uprising intensified, Walker lent his hotel to the British army to use as a lookout point. It was a costly decision. In May 1954, rebel fighters torched the giant fig tree and destroyed the hotel. A bull elephant later pushed what remained of the charred tree to the ground.

Another mugumo a few kilometres away served Dedan Kimathi. This giant's aerial roots had coalesced into three great pillars whose surfaces bore the scars of their birth, myriad cracks and crevices where the roots had not fully merged. Kimathi saw potential in those hollows. The tree's false trunks stretched skyward like a trio of sinewy limbs that had come together to hold aloft the vast crown of leaves.

This landmark would become a Mau Mau communications hub. Its nooks and crannies served as post office boxes for different Mau Mau groups and sympathetic villagers. They used charcoal or even their own blood to write coded messages on strips of animal hide, which they tucked into their intended recipient's specific recess on the tree. Through the tree, the Mau Mau coordinated their troops, shared military intelligence and requested food supplies from villages. The Kimathi Post Office, as it came to be known, is now a national monument in recognition of the role it played in the struggle for independence.

It was a dirty war. The Mau Mau staged some sickening acts of violence against civilian targets. For their part, the British committed massacres and extra-judicial killings. They forced hundreds of thousands of Kikuyu men and women into concentration camps, and subjected many of them to extreme violence and torture. By 1956, the British forces had killed, captured or converted to their cause nearly all of the Mau Mau fighters. Some now worked for the British, hunting down their former comrades.

There was no bigger prize than Dedan Kimathi. A colonial policeman called Ian Henderson led the chase. He remains a controversial figure. In 2013, Kenya's Truth, Justice and Reconciliation Commission would state: 'The historical record suggests that Henderson committed or aided in the commission of extra-judicial killings of Kenyan freedom fighters, as well as contributed to the construction of a system for repressing democratic opposition through illegal killings and enforced disappearances.'

Henderson's breakthrough in the hunt for Kimathi came when his team found a letter that Kimathi had written to his brother. The letter described a dream in which Kimathi's god Ngai had taken him by the hand and walked him through a 'beautiful forest where there were many red and yellow flowers and big birds with green wings'. Ngai took Kimathi to the biggest fig tree in the forest, a mugumo 'that was like a father of all trees'. Ngai told him, 'This is my house in this forest, and here I will guard you.' Then the fig tree rose out of the ground and disappeared into the clouds. When Kimathi woke he could no longer recall where he had seen the tree but knew he must find it. In recounting this dream, Kimathi invited a nightmare.

Henderson's team later caught and interrogated one of Kimathi's lieutenants who revealed that Kimathi made weekly pilgrimages to pray at giant fig trees in the forest. Kimathi believed that if he did this, his god would never let him die. Henderson said he jumped for joy when he learned this. 'We would watch the mugumo trees like hungry vultures and take him by surprise when he came to pray,' he later wrote. 'It would certainly be easier to watch the trees than work almost blindly in those hundreds of square miles of forest.'

'Operation Wild Fig' was underway. Kimathi's lieutenant told Henderson there were at least 40 mugumo trees in the forest where Kimathi hid. Henderson's team found only 18, but could rule out 10 of them in places Kimathi would never feel safe enough to visit. Eight teams of men—a mix of British soldiers, Kenyan Home Guard and Mau Mau insurgents-turned-collaborators—staked out the eight remaining fig trees. They had to avoid leaving a footprint or making any noise, even as horseflies bit at their flesh while they lay in wait. The ambushers watched as butterflies flocked to the bases of two of the trees. They had come to feed on the honey that Kimathi had offered as a libation to Ngai just days earlier. The plan was working.

After three days and three nights, a figure appeared near one of the trees. It was one of Kimathi's men. He had come to see if the tree was safe for his leader to approach. Henderson's team followed his tracks back through the forest for more than 13 miles when Kimathi's voice called out. Within seconds gunfire blazed and a grenade exploded. In the chaos Kimathi escaped. The mission had failed, but it had left Kimathi trapped in a small area of forest. Within weeks, he was caught and in custody. On November 19th, 1956, he was convicted of carrying a firearm without a permit and sentenced to death by hanging.

Legend has it that when Kimathi was executed, on February 18th, 1957, his favourite mugumo prayer tree crashed to the ground. His death marked the end of the uprising. Thirty-two European civilians and 200 members of the security forces had lost their lives. This toll was dwarfed by that of insurgents, thousands of whom were dead. Many Kenyan civilians were caught in the conflict; both side's bullets drew their blood. Britain had quashed the uprising, but its grip over the Kenya colony had become untenable. Within a few years, the British had handed power to African hands, and an old prophecy about another famous fig tree had come true.

In the late 1800s, a Kikuyu seer called Cege wa Kibiru foresaw the arrival of pale-skinned people who would carry 'fire sticks', their guns. He saw an iron snake that would eat people and vomit them out—the train. He also predicted that when a huge fig tree in Thika fell, his people would be free. When representatives of the British Colonial Government heard this story, they reinforced the tree with a fence. It did not help. Part of the tree fell in May 1963 and a month later Kenya had gained internal self-rule. The remainder of the tree fell six months later. Within a month—on December 12th, 1963—Kenya became an independent country, and Jomo Kenyatta its first Prime Minister.

One of Kenyatta's first acts was to plant a mugumo fig tree at the spot in central Nairobi where for years a tall pole had held aloft the British flag. It was a potent statement. Across the world, other colonies that gained their independence adopted fig trees as symbols of their societies. Independent Indonesia had put a fig tree on its coat of arms, as would Barbados. India chose as its national tree the banyan *Ficus benghalensis*, and Sri Lanka would put four *Ficus religiosa* leaves on its national flag. Attempts to reconnect formally colonised cultures with key features of their local ecology were not always matched by policies.

In Kenya, one of the first people to notice and try to correct this was Wangari Maathai, the woman who had played beneath a sacred fig tree as a child.

Three years before independence, Maathai had left to study in the United States, where she picked up bachelor's and master's degrees in biology. Back in Kenya, she would become the first East African woman to be awarded a doctorate. She was not content to be a quiet scientist. When she returned to an independent Kenya in 1966, she found that the giant fig tree she had played beneath as a child had been felled. The stream that ran beneath it had dried up. Maathai was starting to understand why fig trees had become embedded in her culture in the first place. She would prove visionary in her realisation of what it meant for the traditions and the trees to disappear.

As the plantations spread in the 1960s and 1970s, yet more fig trees fell. The forests shrank. Women there told Maathai they could no longer find firewood to cook with, that their soil was eroded and their springs had run dry. When Maathai heard this, she had what she called a 'simple and big idea. . . . It just came to me. Why not plant trees?' In 1977, working through the National Council of Women of Kenya, she set up the Green Belt Movement. Her simple idea was to empower women to plant seedlings of native trees and re-green the denuded land. She made powerful enemies as a result. Politicians and businessmen saw the women as a challenge to their control of the land and of the women themselves.

Maathai's husband divorced her, stating that she was 'too educated, too strong, too successful, too stubborn and too hard to control'. Her biggest challenges were yet to come. She would endure death threats, police harassment, imprisonment and beatings. One left her in a coma. President Daniel arap Moi called her a 'madwoman'. Maathai did not bend. She and her colleagues dug in to score spectacular victories.

In 1989, they blocked a major development President Moi had planned in Nairobi's Uhuru Park. Ten years later, their protests prevented the government from privatising parts of a forest and handing it to cronies. By December 2002, Moi's time was over. Voters kicked out his party, which had enjoyed nearly 40 years of power. Maathai was elected to parliament with 98 per cent of her constituency's vote. She noted that the police who had jailed her the previous year were now her escorts.

For all she endured and all she achieved, Maathai would win the 2004 Nobel Peace Prize. In 30 years, her movement planted tens of millions of trees in Kenya and has expanded to other countries. She showed that something as simple as a tree could provide security, prosperity and hope. Through media interviews and speeches, in books and documentaries, Maathai told millions of people how she traced the roots of her environmental consciousness back to the giant fig tree she played beneath as a child.

Maathai concluded that her people's protection of certain *Ficus* species had more than a religious basis, that it drew also on an ancient appreciation of the benefits these trees provide. Those benefits are many, varied and free. A mugumo's dense canopy provides shade and slows the flow of rainwater onto the ground, limiting soil erosion. Its figs sustain a myriad of birds and mammals, which disperse the seeds of many other species. Its roots are strong; they help stabilise soil and prevent landslides. These roots are also long. When they reach underground water, the liquid can rise up through the channels the roots have carved and break through to the open air to form springs of clean water, even in arid lands. That, said Maathai, is why the figs became trees of God.

Not everyone agreed. This became clear, in 1996, when Kenya's parliament debated whether to protect three mugumo fig trees that had long been sacred to Kikuyu people. 'All that we are asking the Assistant Minister to do,' said Stephen Ndicho, the MP for Juja, 'is to allocate about an acre of land for a mugumo tree and declare it a national shrine, so that if we fail to get our prayers answered in the churches, we can go back to ask our Ngai to give us rain.' 'This is devil worship!' replied Dr Lwali Oyondi, MP for Nakuru. When the Speaker of the House ordered Oyondi to withdraw his comment and apologise, he did only the former.

After centuries of being central to Kenyan cultures and histories, the fig trees are being shuffled off stage. There is a chance, however, that they could return to the limelight. In Kenya, as in many other places, fig trees can be vital allies in our efforts to restore damaged forests and protect wild species. For, as violent volcanoes have taught us, these trees can help restore life to even apocalyptic landscapes.

ELEVEN

The Testimony of Volcanoes

The Time of Darkness was a time of terror. For three days, thick ash choked the sky. Red-hot rocks rained down. They crushed houses and sparked fires where they fell. The debris that plummeted from the air killed many people. Yet more starved to death because the ash fell like hot snow, smothering crops and ruining them. People tell this tale across a vast area of Papua New Guinea, from the Pacific coast to the highlands over 400 kilometres away. Dozens of versions of the story exist in more than 30 local languages. These stories have passed from generation to generation by word of mouth, and although they vary from place to place, on most of the details above they agree.

When geologist Russell J Blong first heard these tales in the late 1970s, he set out to identify the cause of the catastrophe. Clues sprang from the versions of the story that say the darkness followed explosions and earthquakes and immense waves that tore into the shore. To Blong, these were tell-tale signs of a massive volcanic eruption. His research showed the stories originated when Long Island, a volcano 55 kilometres off Papua New Guinea's northeast coast, erupted with titanic fury sometime around 1660.

Long Island is a misnomer. Seen from the air, the island is a roughly hexagonal ring of land surrounding a 13-kilometre-wide lake, which formed in the crater the eruption created. The island got its name in 1700 when British navigator Captain William Dampier sailed past on his way to Australia and saw 'a long island with a high hill at each end; this I named Long Island'. Had Dampier visited a few decades earlier, he would have seen a very different silhouette on the horizon. The huge eruption had decapitated the island.

Dampier wrote that the island 'appeared very pleasant, having spots of green savannahs mixed among the woodland: the trees appeared very green and flourishing, and some of them looked white and full of blossoms'. What he saw was a landscape in recovery from a destructive past. Had he gone ashore he would have found an abundance of fig trees, the vanguard of nature's relentless drive to reclaim and restart.

In the past 10,000 years, there have been only 64 eruptions as big as Long Island's. The Smithsonian Institution's global volcanism programme ranks the eruption as 'colossal' on its 'volcanic explosivity index'. It shot enough material into the sky to fill 12,000 Olympic-sized swimming pools. Some of this matter reached 25 kilometres into the atmosphere. The debris that rained down covered 87,000 square

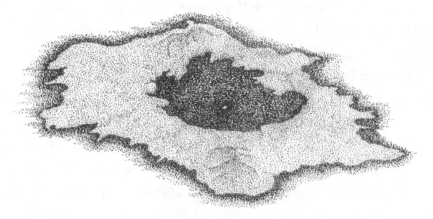

VOLCANO IN RECOVERY:
Fig trees are dominant forces on Papua New Guinea's Long Island and, near
the centre of its crater lake, the tiny island of Motmot

kilometres of New Guinea to a depth of 1.5 centimetres. But on what was left of Long Island itself the blanket of ash was 30 metres thick.

The volcano had sterilised itself. The sole survivor, according to some accounts of the Time of Darkness, was a woman who fled the carnage on a raft and washed up on the mainland. By purging the island of all life, the eruption created a natural experiment that scientists would study centuries later. My call to join that endeavour came in 1999, when, during my research in Borneo, I was invited on an expedition to check on Long Island's recovery. The plan was to spend two weeks camped in the forest there while we identified every plant and animal we could find.

And so, on a June day that year, I stood at the epicentre of the cataclysm, on an outcrop of lava called Motmot in the crater lake of Long Island's still active volcano. It was a fierce, bleak landscape. There was no shelter from the intense sun; the sharp black lava tore shreds from my shoes. This island within an island was 200–300 metres wide and rose 50 metres above the lake's surface. It was mostly barren, for Motmot was young. It had formed in 1968 when a small eruption forced material above the lake's surface. Further eruptions on Motmot in 1973 and 1974 included lava flows that ensured the island's permanence but also wiped out what little life there was there.

By the time I visited Motmot a quarter of a century later, the animal life was dominated by just two species of spider and one kind of ant. But standing there in that hostile envi-ronment was a two-metre-tall *Ficus benjamina* fig tree, the same species whose leaves I had dusted as a child. Other figs were there too. Motmot had just 45 plant species and most were small weedy ferns, sedges and grasses, but among them were eight species of fig trees. Their presence was a testament to the power of these plants to colonise near-lifeless landscapes.

I had gone to Long Island on an expedition led by Ian Thornton, a 72-year-old Yorkshireman who was a professor of biology at La Trobe University in Melbourne, Australia. Thornton was a no-nonsense straight talker with an untamed rebel streak, a wicked sense of humour, and an even stronger sense of justice. When Professor John Zachary Young of University College London criticised the work of one of Thornton's students, Thornton challenged him to a duel. The pair faced off on a table top with toy swords in their hands and did battle until they both fell from the table. When our team met for the first time on mainland Papua New Guinea, Thornton told us he had been hospitalised for a 'slight heart attack' just two days earlier. He said he would be 'taking it easy'. He then took a glug of cold beer and dragged on a Spear—a six-inch-long local cigarette. We were in for an adventure.

Within 48 hours, Thornton and another professor, John Edwards of the University of Washington, would be stranded on Motmot, their inflatable boat having slipped its mooring and scudded away across the lake. Two other colleagues were missing, adrift somewhere in the water having tried in vain to swim after the boat. Thanks to some quick thinking, clear protocols and a dose of good luck, everyone was safe by sundown. The following night, the malaria pills I was taking induced me to leap up and charge into the forest where I collapsed, hallucinating and convinced I could not breathe. On later nights, men from the nearest village snuck into our camp on the shore of the crater lake to steal clothes, shoes and food. We ran out of supplies and were stranded on the island when our return boat failed to appear. We eventually made our way back to the mainland on a small fishing boat. Dolphins escorted us, riding the boat's bow wave as we filled our hungry bellies with pan-fried tuna that had been swimming in the sea just moments earlier.

In between these episodes, the biology kept us busy. We found that at least 50 species of land birds had colonised the island since it exploded, as well as thirteen species of mammals, fourteen reptiles and two amphibians. More than half of the bird and mammal species were fig-eaters. Their home was a forest in which fig trees were a dominant force. We found 31 species of *Ficus* there. That's more than one in ten of all of Long Island's plant species. Some of the *Ficus* species may have arrived when people resettled Long Island and brought seeds and plants they could grow for food. Most though would have arrived as seeds carried by fruit bats and pigeons that had fed on figs on the mainland before flying over 45 kilometres of open sea to Long Island.

What's more remarkable is that in just two weeks we found 16 of these *Ficus* species with figs on their branches. For each of them, the pollinator wasps had successfully colonised Long Island too. Our snapshot survey revealed the power of fig trees to sustain the island's fruit-eating animals. Some of the *Ficus* species had large green figs, which would attract bats but fail to arouse any interest among the birds. Others had small red figs that would turn the tree's crown into a carnival of colour as yellow-billed fruit doves, rainbow lorikeets and metallic starlings flocked to feed.

These animals and many others would struggle to survive on Long Island without its year-round supply of figs. For such species to establish themselves after the island's eruption, the fig trees must have been there first. That's why the fig trees on Motmot were so interesting. They showed that *Ficus* species would have been able to colonise Long Island itself fairly soon after it erupted, their seeds germinating in little more than bare lava, having been dispersed by birds and bats roaming in search of food.

Ficus species are better colonists than most trees because they have small seeds that fig-eating animals can carry long distances in their guts before they excrete them. Fig trees produce these seeds in great numbers, more than once a year, and this raises the odds of some of them reaching new land. Once there, the seeds that germinate produce roots that grow fast and strong, even in the toughest of ground, and can keep growing through or over several metres of earth until they reach water.

Of course, for a population of any *Ficus* species to thrive, it needs its pollinator wasps, but they too can travel tens of kilometres in just the day or two they live. Fig-wasps have

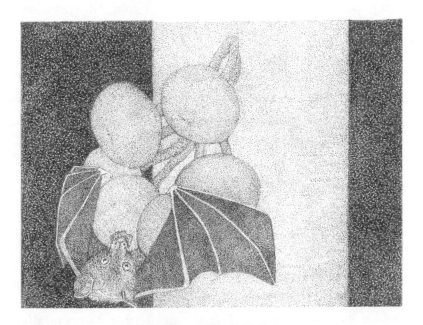

NIGHT GARDENERS:
Fruit bats have been important dispersers of fig seeds for tens of
millions of years

been caught in traps on ships up to 50 kilometres offshore. Research in Namibia has revealed that fig-wasps can carry pollen more than three times that distance.

Colonisation is only the start of a fig tree's power. When pioneer *Ficus* trees are big enough to produce figs and their wasp partners pollinate the flowers within, the resulting crops of ripe figs attract fruit-eating animals, which carry the seeds of other plant species in their guts. It is likely that on Long Island, fig trees played a critical role in giving seed-dispersing animals a reason to visit. As the pioneer figs grow, they provide important shade for other species. The fig trees herald the arrival of many more plant species, and so a forest forms.

Researchers have measured this effect in a region of central Mexico called Los Tuxtlas where big *Ficus* trees stand isolated in fields—lone remnants of former rainforest. When Sergio Guevara and colleagues collected all the seeds that fell beneath five of these isolated fig trees, they identified 149 different plant species, mostly those dispersed by fruit-eating birds and bats. The researchers also recorded 47 species of fruit-eating birds visiting the trees. They erected fences around the trees to keep cattle out, and after three years, a dense layer of new trees of dozens of species, some four to five metres tall, had established themselves there. Guevara's team showed that even a single fig tree can have a profound effect on the land that surrounds it. Taken together, such research suggests that when Long Island hit the road to recovery, fig trees were in the driving seat.

Long Island is not alone. Two centuries after it erupted, another natural experiment began with a bang that people felt all around the world. On August 27th, 1883, Krakatoa, a volcanic island west of Java, Indonesia, erupted with such spectacular force that it triggered a tsunami that killed people 3,000 kilometres away in Sri Lanka.

Early in 1884, the French government sent investigators to report on the aftermath of the eruption. They said that, in Bantam, on the Javan mainland 50 kilometres away from Krakatoa, some fig trees were all that remained of the forest: 'The wave which rushed with such force upon this coast destroyed the forest for a distance of three hundred or four hundred metres inland, leaving nothing standing except the great *Ficus religiosa*, which stretched their dry and bark less stems toward the heavens.' On Sebesi, an island 22 kilometres from Krakatoa, 'the destruction was complete— hardly a bit of herbage, hardly a trace of life remained'. Sebesi was buried to a depth of 10 metres in cinders and ash. A large portion of Krakatoa itself had vaporised. On what was left of the island, the ashes were 60–80 metres deep. Everything on Krakatoa and the two islands nearby was dead.

Visit today and you will see a very different picture. In little more than a century since the volcano erupted, more than 200 plant species arrived. Thick forest covers what remains of Krakatoa—from the coast to its 800-metre peak. The animals that have colonised include 30 species of birds, 17 species of bats and thousands of insects and other invertebrates. The wings of more than 50 kinds of butterflies bring flashes of colour to the once lifeless island. Each of these species had to cross at least 44 kilometres of sea to reach the island then establish themselves there for the long term.

As on Long Island, fig trees appear to have been instrumental in establishing nature's toehold on the land and attracting other species to follow suit. In fact several of the same *Ficus* species are present on both of these islands, despite them being more than 4,500 kilometres apart. These volcanoes highlight the ways in which different kinds of fig trees have different biological powers. Whilst it is the giant trees and strangler figs that offer most fruit to animals in mature forests, it is often the smaller species of fig trees that are best at colonising bare land and kick-starting rainforest regeneration.

Long Island and Krakatoa show that, if left alone for long enough, rainforests can recover from even catastrophic damage. But they also show that there is an order to all things in nature, and that before big rainforest trees must come smaller pioneers that attract animals that disperse the seeds of other species. The rebirth of life on these tropical volcanoes suggests that fig trees could help forests to recover elsewhere in the tropics, where logging and mining have taken their toll. Nature, though, may be too slow to withstand what we throw at it. It may need a helping hand.

Around the world, scientists are using fig trees to provide that help. One project in Indonesia is looking at establishing strangler figs to boost seed disperser populations in a protected area in the hope they will fly out over degraded forests and spread seeds there. In another project, in Rwanda, researchers have planted 400 branches from mature *Ficus* species to assess whether these trees can help restore forest rapidly. In Costa Rica, biologist Rakan Zahawi scaled up this approach, lopping off huge branches, more than four metres long, from *Ficus pertusa* trees and planting the stakes as 'instant trees'.

In each of these projects, the aim is to increase the availability of figs and so attract and sustain populations of animals that disperse seeds of other forest trees. Birds, primates, fruit bats and other animals do indeed provide a free dispersal service. But as the recent history of forests worldwide shows, when the bigger birds and mammals disappear, so do the services they provide. When biologist Steve Elliott got to thinking about how to overcome this challenge, he found an answer, or rather a question: Can robots restore rainforests?

TWELVE

Once Destroyed, Forever Lost?

'We were viewed as crackpots even by conservationists. They thought we were mad. We got opposition from just about everyone.' It's 2013, and Thailand-based biologist Steve Elliott is telling me a remarkable story in which fig trees are the heroes. Today the doubters are converts. Elliott wants them to follow him into a future where technology and ecology combine to restore degraded rainforests.

Elliott ended up working in Thailand by accident. He went there on holiday on his way back to his native Britain after completing his PhD research on medicinal plants in Indonesia in 1986. Before long, three Thai universities had offered him jobs. He chose Chiang Mai University in the north of the country, where a one-year contract to teach wildlife conservation would grow into a career. It was an ecologist's dream—the university sits right on the border of Doi Suthep-Pui National Park.

'You could finish a lecture at 11.30,' he says, 'and by

12.30 be sitting on a log, eating sandwiches for lunch, in a fairly pristine natural forest.' The park covers steep mountain slopes and is home to more than 300 species of birds. But despite the high biodiversity, large areas of the park were treeless. Local Hmong villagers had converted nearly a fifth of it to farmland. Cabbages, carrots and corn now grew where tall trees once stood. The villagers had abandoned some of their fields, and the weeds that now choked them prevented the forest's return.

One community that lived in the national park's Mae Sa valley learned the hard way what the loss of trees can mean. In the 1960s, the villagers felled most of the surrounding forest for timber to build houses and for fuel wood to cook food. But with the trees gone, the spring the villagers depended upon dwindled to a trickle then died. Without water, the people abandoned their homes and moved downhill to start again, but they did not forget the lesson nature had taught them. Three springs flowed in the forest that surrounded their new village of Baan Mae Sa Mai. The villagers protected this area as a community forest, and with it the water that bathed their babies and fed their fields downstream. In the mid-1980s, they set up a conservation group and began to plant trees in the upper watershed. But their good intentions were not matched by results.

The Royal Forest Department of Thailand had given the villagers seedlings of exotic trees such as eucalyptus to plant, but villagers did not like them. They burnt easily and were good neither for local wildlife nor to protect the watershed. 'When we said we wanted to test native trees, they welcomed us with open arms,' says Elliott. He had no idea he would be working in the village for 20 years. That was in 1996. Two years earlier, Elliott and one of his university colleagues, Vilaiwan Anusarnsunthorn had founded the Forest Restoration Research Unit (FORRU) to work out how to restore forest on land that had been logged, farmed and abandoned.

FORRU convinced the Doi Suthep-Pui National Park authorities to support their plan to learn how to grow native trees. The FORRU team gathered any seeds they could find in the forest. They picked them up from the ground. They climbed trees to get them. They used 10-metre-long poles to knock down fruit. Corner's botanical monkeys would have come in handy. The plan was to germinate the seeds in a nursery, let the seedlings grow for a year then plant the young trees in a deforested area of the National Park. But the first experiments failed. 'We started planting out trees from the nursery,' says Elliott, 'and then pretty much monitored them as they died.'

It became clear that very few species would survive without considerable and costly help in the form of manual weeding and fertiliser. FORRU was floundering, but everything changed when, later in 1996, Elliott went to a conference in Washington, DC. It was one of the first big international gatherings of experts in the art of tropical forest restoration. The conclave covered a wide spectrum of approaches, from eucalyptus plantations to agroforestry and the restoration of natural forest ecosystems, but the speaker who had the greatest impact on Elliott was an Australian called Nigel Tucker.

Tucker had grown up in the vast outdoors of north Queensland where, aged just 15, he discovered a species of legless lizard that was new to science. The wilderness was both his playground and his teacher. It would become his workplace. As an adult, he worked to restore the degraded forests there, to reconnect patches of habitat and heal damaged land. Along the way, he developed what would become known as the framework species approach. This involves planting species that can shade out weeds and produce fleshy fruit that attract seed-dispersing animals from nearby patches of intact forest. These animals bring in their bellies the seeds of other plant species and then deposit them when they defecate. The seeds are better able to germinate and thrive in the weed-free shade that the planted trees provide. From the ground up, the forest's physical structure and species composition return.

Tucker, in turn, recalls Elliott's talk about his challenges in Thailand well. 'His presentation in Washington took me back 10 years,' he told me. 'Obviously, you don't want to see people repeat the painful lessons you've already learned, so over pizza one night I suggested he and his coresearcher, Kate Hardwick, come to north Queensland and have a look at our techniques.' Elliott's visit to see Tucker's framework species demonstration plots in early 1997 would knock 10 years off FORRU's struggling research programme. That's when Elliott learned the power of the fig trees. 'What we had been doing was reinventing the wheel, going through the same processes Nigel had struggled with years earlier,' says Elliott. 'One of the first things he told us was that nearly all *Ficus* species can act as framework species. Provided you select species that grow naturally in the type of forest ecosystem you are trying to restore, then you can plant them and they will boost ecosystem recovery.'

Back in Thailand, the FORRU team ploughed through their data and identified local trees, including several kinds of *Ficus*, with the same characteristics as Tucker's framework species. By June 1997, they were ready to take their new tools into the forest, and the village of Baan Mae Sa Mai. In just a year the results were clear. 'The villagers were scratching their heads,' says Elliott. 'These trees were waist high. Normally the eucalyptus would be dying, and the pines covered in weeds. We went from watching trees die slowly to watching them grow fast and close canopy in two years.' But it was just a start. Within eight years of FORRU planting their framework species, more than 70 other tree species had recolonised the experimental plots.

The *Ficus* species played a critical role, thanks to a set of traits that make them so suited to the framework-species method. First, they have phenomenal roots, which grow fast and can even tear rocks apart. 'They can find water in the dry season when other tree species can't,' says Elliott. 'At the end of the rainy season, there are three to four months without rain. Fig trees find so much water with their fantastic roots that they are evergreen when all around them, in lowland areas of the park, trees have shed their leaves.' Second, *Ficus* species grow very fast and their thick green leaves cast a dense shadow and shade out grasses and climbers. Third, fig trees are magnets for biodiversity, attracting animals that disperse the seeds of many other species—including of course more fig trees. Within just three years, several of the *Ficus* species FORRU planted produced figs that attracted birds and mammals such as monkeys, civets and barking deer. Other planted fig trees served as nest sites for seed-dispersing birds. The number of bird species rose from 30, before planting, to at least 87.

Tucker says at least one in every five seedlings planted should be a fig species. The FORRU team followed this advice and used experiment after experiment to refine the knowledge they needed to ensure that seeds become trees, and that trees grown in the nursery survive when someone takes them into the forest and plants them in the ground. They worked out which containers and potting medium worked best, as well as how and when to water and fertilise the seedlings in the nursery. Their forest grew. By 2008, they would write that FORRU 'rejects the adage, adopted by many conservation organisations, that "tropical forests, once destroyed, are lost forever". The unit bases its work on the more optimistic view that it is possible to transform largely deforested landscapes into lush tropical forests in a few years.'

They are mimicking what happened on the volcanoes of Krakatoa, Indonesia and Papua New Guinea's Long Island, where fig trees helped forests return to what was once black lava. But they are doing it much faster, at a cost of about one dollar for each tree established from seed. The villagers have benefited too, and from more than just the jobs the project created. More trees means more biodiversity, more secure water supplies and less risk of flooding or landslides or soil erosion. And by showing themselves to be effective custodians of the natural resources that surround them, the villagers stake a claim to continue living in the national park.

The FORRU project has inspired several international organisations to implement similar framework-species schemes elsewhere in Asia and in Africa. Buoyed by their success, the FORRU team now wants to set fig trees a tougher test—to restore landscapes that seem to be beyond repair: the rocky, churned-up scars left behind by open cast mining. It comes from the observation that fig seeds germinate just about anywhere. 'All over the university campus, we see fig seeds germinating in cracks in walls and walkways and as the roots of the trees expand, the cracks get wider,' says Elliott. 'If they can do that to university infrastructure, they will have no trouble opening up mine substrates, but the trick is to encourage them to germinate under such harsh conditions.'

The hope is that fig tree roots will break open the rocks, create drainage channels and allow oxygen to enter the substrate. As soil begins to form, these conditions will enable other, less hardy, tree species to colonise the site. Local wildlife would disperse seeds from these other species when they come to eat figs.

Today, says Elliott, FORRU is overwhelmed with interest. The turning point came at the UN climate change negotiations in 2007. That's when nearly 200 governments began to work out how to compensate nations that protect forests, locking away carbon that would otherwise contribute to climate change. The original proposal focused on avoiding deforestation to prevent carbon entering the atmosphere, but after years of intergovernmental negotiations, it now includes 'enhancement of carbon stocks'—in other words, forest restoration—and aims also to preserve biodiversity and benefit local communities.

It just so happens that FORRU has developed a way to do all three. But the challenge is scale, as fig biologist Rhett Harrison points out. 'Planting out large numbers of seedlings, even with the relatively efficient framework approach, is expensive and not really an option if you are faced with hundreds of thousands or even millions of hectares to restore.'

Elliott recognises this. He says while the framework species approach seems to work and villagers are keen to use it, it is a big job to collect seeds year-round, grow trees for two years in a nursery and then carry them to distant and often steep areas to plant them. The villagers would have to put the trees in baskets and haul them up on foot, often walking for hours with 20–30 kilograms on their backs. As Elliott says: 'People don't want to lug seedlings down 45-degree slopes.' His plan is to fly unmanned aerial vehicles—drones—over remote sites and drop fig seeds in containers of hydrating gel. Tests are already underway. Ultimately, Elliott sees a role for robots in seed collection too.

In stage one of his vision, people would use smartphones to control small flying robots with rotor blades at each of their four upper corners. These highly manoeuvrable drones could fly deep into a forest. A camera mounted upon them would transmit video back to the smartphone and allow the controller to locate fruiting trees. The robots would also transmit the GPS coordinates of such trees, making it simple for people to find them to collect the seeds.

In stage two, says Elliott, the robots would take over. They would use image-recognition software to identify tree species and electronic tools to harvest and carry their fruit. They would return each day to their base, a bamboo hut mounted with solar panels, to recharge their batteries by landing on electromagnetic induction pads. 'After recharging overnight,' says Elliott, 'our little flying robots are ready to continue their mission—to seek out new fruiting trees— to boldly go where no one has collected seeds before.' Elliott says the only human intervention needed would be for someone to pick up the fruits collected by the flying robots and perhaps clean off the solar panels and repair any mechanical failures.

'Parts of the vision are already possible,' says Lian Pin Koh, professor of applied ecology and conservation at the University of Adelaide's Environment Institute. 'For example, sending a drone on an autonomous mission to take images of the forest canopy to detect flowering or fruiting trees. Currently, the technology still doesn't allow us to send drones under the canopy, although that might change pretty soon. The other parts of the vision, on induction charging

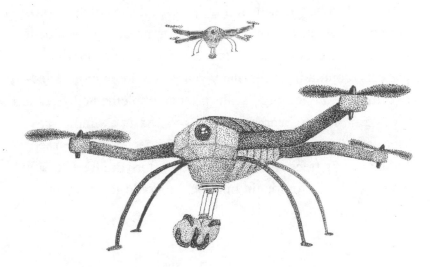

ROBOTS IN THE RAINFOREST:
Aerial drones on a mission to gather rare fig seeds
for other drones to disperse

and image recognition, are probably five to ten years away.' By the time you read this, drones may already hover over forests, doing the work of wildlife that has vanished.

With or without robots, Elliott recommends replicating, over many communities, the small-scale work FORRU has pioneered rather than planning large-scale restoration programmes, which has often been the approach government agencies have adopted so far. He says many small-scale projects, continuing over many years, would be more effective than a single, large-scale project implemented over a few years. At the heart of Elliott's vision is a simple message, one that was among the first our species ever learned: Look after fig trees and they will look after you. It's a lesson we have all but forgotten, but one we could learn again.

EPILOGUE:

A *Wedding Invitation*

The bride and groom were as silent as stones throughout the wedding that took place on June 10th, 2008, in Jimiti, a village in the Indian state of Odisha. This was no ordinary marriage, for they were both fig trees, he a banyan (*Ficus benghalensis*) and she a pipal (*Ficus religiosa*). The villagers spoke for the trees, the men representing the groom and the women the bride.

The practice of marrying trees is ancient and widespread in India; it has various meanings, but this time the purpose was to raise awareness of the need to protect trees. 'Our village was covered with hundreds of trees four decades ago,' said local man Nabin Rout in an interview with *The Times of India*. 'But the green areas have been shrinking year by year, and if we do not protect them they will disappear in a few years.'

Jimiti village is by no means unique. Across the world, forests fall faster than they grow. Deforestation drives climate change by destroying trees that suck heat-trapping carbon out of the atmosphere. Populations of forest creatures shrink in number and range and so are less able to provide services such as seed dispersal and pollination. Extinction claims species before we can even name them. These trends, born of our own success, threaten to diminish our children's futures.

Forests are where we came from and forests retain a strong grip on our species today. But there's a paradox at play. The more forests there are in the world, the more liberated humanity can be. By contrast, the more forests we fell, the more we play dice with the climate, with water supplies, with complex ecological webs that provide us with vital goods and services. What should concern us is not the resilience of nature, but our own resilience to the new nature we are bringing into being.

It will take more than a marriage of fig trees to solve our problems. But other kinds of weddings that involve fig trees could help. I'm thinking of marriages of world views— those of scientists and religious people, of environmentalists and economists, of poets and politicians—because stories of fig trees and what they offer may resonate with these groups in different, yet complementary, ways.

Fig trees are great connecters. Whatever our political or philosophical differences, we are all descended from some fig-eating ancestors. It is a good starting point. Fig trees stood tall long before the first human footprint graced the Earth. They helped to make us and the world we inhabit. The gifts they offer are ancient but remain relevant today.

We can use these trees not only to strengthen our standing forests but also to restore those that have fallen. On logged land, *Ficus* species can kick-start regrowth of rich forests whose trees and soils lock away carbon and so limit global warming. They can provide materials and medicines that turn poor farmers into entrepreneurs. Their figs can sustain the wild birds and mammals whose daily defecations will plant the forests of the future.

To take these gifts, we will need to protect both the trees and the animals with which they are bound by contracts of codependence. We will need to do this in ways that are rooted in science but also owned by, and beneficial to, local communities. In this respect, fig trees also have great potential. They already have prominent roles in many cultures, in some cases forgotten but waiting to be remembered. They are symbols of what we all share, and their success and longevity as a group puts humanity's short stint on Earth into humbling perspective.

Fig trees have watched in silence as civilisations have climbed high only to crumble. From the lost city of Gedi in Kenya and the abandoned town of Kayaköy in Turkey to the slopes of Mohenjo-daro in Pakistan, fig trees have reclaimed landscapes no longer in the grip of human hands. The people who lived in Mohenjo-daro adopted fig trees as symbols more than 4,000 years ago. Life was good until it wasn't. By the time Charles Masson walked there in 1827, there was little to see other than some mysterious mounds from which ruins poked like broken teeth. Masson said the whole area was populated by giant *Ficus religiosa* trees 'some of them in the last stage of lingering existence; bespeaking a great antiquity, when we remember their longevity.' They were living links to a long-dead civilisation.

Local people told Masson there had once been a vast urban centre there but that it 'was destroyed by a particular visitation of Providence, brought down by the lust and crimes of the sovereign'. In fact, it was local climate change and a brutal long-term drought that forced the people to abandon their city. The course of the Indus River, upon which they depended, had shifted and sentenced their civilisation to oblivion.

Far away in present-day Guatemala and Cambodia, droughts hundreds of years ago forced the Maya and Khmer people to abandon fabulous cities that were destined to share something curious in common, something at once beautiful and profound. The surrounding forests swiftly reclaimed the abandoned buildings. In each case awesome strangler figs accelerated the return of nature. Their seeds germinated in cracks in the stonework and grew into giant plants that in time disguised or demolished the stone structures. Their ripe figs attracted hungry animals that brought, in their bellies, the seeds of many other rainforest species.

Centuries later, the ruins have since been unearthed and protected as tourist attractions, but the stranglers remain, their sinuous aerial roots festooned over ruined temples and palaces. Inch by inch these elephant-grey roots tighten their grip on the ancient buildings. They are living reminders that the good times don't last forever, that despite humanity's perceived dominion over all creatures great and small, we remain part of nature, not apart from it.

Fig trees have been on Earth for about 80 million years longer than humans. They have seen off asteroid impacts and climate change that wiped out millions of other species. Their story reminds us that we are just new here and that between our kisses, our fights, our struggles and our smiles, we tend to break things before we realise how much we need them. It's a story that tells us much about where we have come from and where we might go from here.

NATURE VERSUS STRUCTURE:
A giant strangler fig dominates a ruined temple in Cambodia

ACKNOWLEDGEMENTS

I am grateful to the many, many people whose generosity and good ways have enabled me to bring this book into the world. First, I owe huge thanks to everyone who supported the book by pledging through my U.K. publisher, Unbound. Without them it would simply not exist.

Thanks, too, to everyone at Unbound and especially Phil Connor, who saw potential in my idea, gave me the green light to pursue my dream and guided me through the process. Thanks also to Unbounders Emily Shipp, Jimmy Leach, Isobel Frankish, DeAndra Lupu, Mathew Clayton and Georgia Odd for your invaluable advice and support.

I thank Kris McManus for producing such a great pitch video, Paul Fulton for some incisive copyedits, and Mark Ecob for letting me get involved in designing the book's cover. That was a great experience.

I am indebted to the community of fig biologists—too many to name here—whose work has inspired much of this book. I have had the great pleasure of meeting and working with many of these scientists. What they have discovered is way more mind-boggling than anything I have written here. Interested readers can find references to their key works in the list of sources at the end of the book.

I owe special thanks to the four men who were my guides through my years as a fig biologist: Steve Compton, Rhett Harrison, Siba anak Aji and the late Ian Thornton. Your knowledge and ideas have informed and inspired me. Without you I would not have found myself so often in ancient forests, awestruck. I can't thank you enough.

It was Thornton who led the expedition to Long Island described in Chapter 11. That trip was one of my life's best experiences, thanks in large part to the great team Prof assembled: Ruby Yamuna; Rose Singadan; William Boen; Ross Clark; Clint Schipper; Simon Cook; Rhett Harrison and the late John Edwards. I'm grateful for all the wit and the wisdom you shared during that special fortnight.

Many people sent me their research papers, answered my questions, reviewed drafts of my text or otherwise encouraged me to write this book. I thank you all: Alex Kirby; Alfonso Peter Castro; Allen Herre; Asko Parpola; Barbara Kiser; Bas Verschuuren; Ben Knighton; Beverly Natividad; Bruce Beehler; Caspar Henderson; Catherine Brahic; Charles Smith; Charlotte Forfieh; Cornelis Berg; Craig Stanford; David Wilson; Eden Cottee-Jones; Edward Lempinen; Elisabeth Kalko; Emmanuelle Normand; Emily Martin; Finn Kjellberg; Fred Pearce; Henry Howe; James Fahn; Jay Matternes; John K. Corner; John Lonsdale; John

Rashford; John Terborgh; Jonathan S. Walker; Joyce Tyldesley; June Rubis; Karline Janmaat; Kathy Willis; Katrin Böhning-Gaese; Keith Summerville; Kenneth Sayers; Kent Redford; Kimani Chege; Laurie Godfrey; Lian Pin Koh; M. Amirthalingam; M.G. Chandrakanth; Madeleine Nyiratuza; Matthew Muriuki Karangi; Mauricio Anton; Michael Renner; Nanditha Krishna; Nathaniel Dominy; Nigel Tucker; Oliver Heintz; Paul Sillitoe; Perpetua Ipulet; Peter Ashton; Peter Schmidt; Rakan (Zak) Zahawi; Rhett Harrison; Robert Hodgkison; Rosie Sharpe; Russell Blong; Shardul S. Bajikar; Shonil Bhagwat; Siew Te Wong; Silvia Lomascolo; Song Qishi; Stéphanie M. Carrière; Steve Compton; Steve Elliot; Ted Fleming; Tim Denhan; Tim Laman; Tim White; TV Padma; W. Daniel Kissling; Wanjira Mathai and William Murphy.

Thank you to my parents, John and Jennifer Shanahan, for nurturing my curiosity and my desire to travel and learn about the world around us. Finally, I want to thank my wonderful partner, Charlotte, who has inspired and encouraged me throughout the process of bringing this book to life, and who more than once talked me out of giving up. This book is for you and our beautiful boy, Noah.

Sources

CHAPTER ONE: Snakes and Ladders & Tantalising Figs

Arita, HT & Wilson, DE (1987). Long-nosed bats and agaves: The tequila connection. *BATS Magazine*, Winter 1987.

Austin, DF (2004). *Florida Ethnobotany*. CRC Press.

Bauer, G et al. (2013). Investigating the rheological properties of native plant latex. *Journal of the Royal Society Interface* 11, 20130847.

Berg, CC & Corner, EJH (2005). *Ficus*—Moraceae. *Flora Malesiana*, Series I, 17, 1–730.

Berg, CC (1989). Classification and distribution of *Ficus*. *Experientia* 45, 605–611.

Datwyler, SL & Weiblen, GD (2004). On the origin of the fig: phylogenetic relationships of Moraceae from ndhF sequences. *American Journal of Botany* 91, 767–777.

Gandhi, M (2009). Fresh figs subzi. Kitchen Tantra—tease your palate. A food blog for Indian recipes, 19 February 2009.

Homer, *The Iliad*, trans. I Johnston (2006). Richer Resources Publications.

Homer, *The Odyssey*, trans. I Johnston (2006). Richer Resources Publications.

Sources

Shanahan, M et al. (2001). Fig-eating by vertebrate frugivores:
a global review. *Biological Reviews* 76, 529–572.

Simoons, FJ (1998). *Plants of Life, Plants of Death*. Diane
Publishing Co.

Wilson, D & Wilson, A (2013). Figs as a global spiritual and
material resource for humans. *Human Ecology* 41, 459–464.

CHAPTER TWO: Trees of Life, Trees of Knowledge

'JM' (1905). A man of the time: Dr Alfred Russel Wallace and his
coming autobiography. *The Book Monthly*, May 1905.

Anon. (1911). We are guarded by spirits, declares Dr AR Wallace.
New York Times Magazine, 8 October 2011.

Beccaloni, G & Smith, C (2015). Biography of Wallace [online
article]. www.wallacefund.info.

Berry, A (2013). Alfred Russel Wallace—natural selection, socialism,
and spiritualism. *Current Biology* 23, R1066–R1069.

Bhikkhu Bodhi (2000). *The Connected Discourses of the Buddha: A
New Translation of the Samyutta Nikaya*. Wisdom Publications.

Chen, A (2005). Weight of expectations proves too much for the
'jinxed' Wishing Tree. *South China Morning Post*, 13 February
2005.

Coder, KD (1996). Trees and Humankind: Cultural and Psychological
Bindings. University of Georgia Cooperative Extension Service
Forest Resources Unit Publication FOR96–46, 10.

Cumont, FVM (2008). *Mysteries of Mithra*. Cosimo Inc.

Gandhi, M & Singh, Y (1990). *Brahma's Hair: The Mythology of
Indian Plants*. Rupa & Co.

Grenand, F (1982). Et l'homme devint jaguar: Univers imaginaire et
quotidien des indiens wayapi de Guyane. *Collection Ams indiens*.
l'Harmattan.

Harding, L (2006). *Holy Bingo, the Lingo of Eden, Jumpin' Jehosophat and the Land of Nod: A Dictionary of the Names, Expressions and Folklore of Christianity*. McFarland.

Hooper, R (2015). When nature evolves to be awesome. *The Japan Times*, 14 March 2015.

Keltner, D & Haidt, J (2003). Approaching awe, a moral, spiritual, and aesthetic emotion. *Cognition and Emotion* 17, 297–314.

Lam, A (2005). Offerings ban on Tai Po Wishing Tree to stay. *South China Morning Post*, 23 February 2005.

Lurker, M (2004). *The Routledge Dictionary of Gods and Goddesses, Devils and Demons*. Routledge.

Martratt, JI (2008). What is taotaomona tree? *Pacific Edge*, 17 October 2008.

Mwakikagile, G (2007). *Kenya: Identity of a Nation*. New Africa Press.

Peters, S (2011). *Material Revolution: Sustainable and Multi-Purpose Materials for Design and Architecture*. Walter de Gruyter.

Piff, P & Keltner, D (2015). Why do we experience awe? *New York Times*, 24 May 2015.

Ramirez, BW (1977). Evolution of the strangling habit in *Ficus* L. subgenus *Urostigma* (Moraceae). *Brenesia* 12/13, 11–19.

Rashford, JA (2013). Candomblé's cosmic tree and Brazil's *Ficus* species, in R Voeks & J Rashford (eds.) *African Ethnobotany in the Americas*. Springer, 311–333.

Simoons, FJ (1998). *Plants of Life, Plants of Death*. Diane Publishing Co.

Slenes, RW (2007). L'arbre Nsanda replanté: Cultures d'affliction Kongo et identité des esclaves de plantation dans le Brésil du Sud-Est (1810–1888). *Cahiers du Brésil comtemporain* 67–68, 217–313.

Smith, WR (1932). *Myths and Legends of the Australian Aboriginals*. Farrar & Rinehart.

Sources

Stoeltje, BJ (1995). Asante queenmothers: a study in identity and continuity, in M Reh & G Ludwar-Ene (eds.) *Gender and Identity in Africa*. LIT Verlag, 15-32.

Tai, E (2000). The Wishing Tree. *Varsity* magazine (Chinese University of Hong Kong), March 2000.

van Wyhe, J & Rookmaaker, K (2013). *Alfred Russel Wallace: Letters from the Malay Archipelago*. Oxford University Press.

Wallace, AR (1858). On the tendency of varieties to depart indefinitely from the original type. *Proceedings of the Linnean Society of London* 3, 53–62.

Wallace, AR (1869). *The Malay Archipelago: The Land of the Orang-Utan, and the Bird of Paradise*. Macmillan.

Wallace, AR (1878). Tropical vegetation. *Tropical Nature and Other Essays*. Macmillan and Co., 27–68.

Wallace, AR (1885). Are the phenomena of spiritualism in harmony with science? *The Sunday Herald* (Boston), 26 April 1885.

Wallace, AR (1889). *A Narrative of Travels on the Amazon and Rio Negro*. Reeve and Co.

Wood, J (2013). Wallace as a writer. *Current Biology* 23, R1072–R1073.

CHAPTER THREE: A Long Seduction

Cunningham, A (1875). Harappa. *Archaeological Survey of India Report for the Year 1872–3* Vol. 5, 105–108.

Farmer, S (2004). Mythological functions of Indus inscriptions. Sixth Harvard Indology Roundtable, 8–10 May 2004.

Forlong, JGR (1883). *Rivers of Life. Or Sources and Streams of the Faiths of Man in all Lands*. Vols 1 and 2. Bernard Quaritch.

Foster, P (2007). Film star faces lawsuits after marrying a tree. *Daily Telegraph*, 1 February 2007.

Frazer, JG (1890). *The Golden Bough: A Study in Comparative Religion*. Macmillan.

Galil, J (2008). *Ficus religiosa* L.—the tree-splitter. *Botanical Journal of the Linnean Society* 88, 185–203.

Geiger, W (1912). *The Mahavamsa. The Great Chronicle of Lanka*. (Translated from Pali). Ceylon Government Information Department.

Griffith, RTH (trans.) (1896). *Hymns of the Atharva Veda*. E. J. Lazarus & Co.

Griffith, RTH (trans.) (1896). *Hymns of the Rig Veda*. E. J. Lazarus & Co.

Haberman, D (2013). *People Trees: Worship of Trees in Northern India*. Oxford University Press.

Krishna, N & Amirthalingam, M (2014). *Sacred Plants of India*. Penguin.

Mansberger, J (1987). In search of the tree spirit: evolution of the sacred tree (*Ficus religiosa*). MA thesis. University of Hawaii.

Parpola, A (1988). Religion reflected in the iconic signs of the Indus script: penetrating into long-forgotten picto+graphic messages. *Visible Religion* 6, 114–133.

Parpola, A (2005). Study of the Indus script. *Transactions of the International Conference of Eastern Studies* 50, 28–66.

Parpola, A (2009). 'Hind Leg' + 'Fish': towards further understanding of the Indus Script. *Scripta* 1, 37–76.

Parpola, A (2010). A Dravidian solution to the Indus script problem. Kalaignar M Karunanidhi Classical Tamil Research Endowment Lecture. World Classical Tamil Conference, Coimbatore (25 June 2010).

Parpola, A (2015). *The Roots of Hinduism: The Early Aryans and the Indus Civilization*. Oxford University Press.

Sargeant, W (trans.) (2009). *The Bhagavad Gita: Twenty-fifth Anniversary Edition*. Excelsior editions, State University of New York Press.

Sources

Seneviratna, A (1994). *King Asoka and Buddhism: Historical and Literary Studies.* Buddhist Publication Society.

Shanahan, M et al. (2001). Fig-eating by vertebrate frugivores: a global review. *Biological Reviews* 76, 529–572.

Strong, JS. *The Legend of King Aśoka: A Study and Translation of the Aśokāvadāna.* Motilal Banarsidass.

Tewary, A (2013). Poor Indian children make a living from 'holy leaves' [online article]. BBC Online, 9 March 2013.

CHAPTER FOUR: Banyans and the Birth of Botany

Agarwal, P (2015). Municipal Corporation to light up memorial of 257 freedom fighters. *The Times of India*, 16 February 2015.

Arrian. *Indica*, trans. PA Brunt (1983). Loeb Classical Library.

Bodson, L (1991). Alexander the Great and the scientific exploration of the oriental part of his empire. An overview of the background, trends and results. *Ancient Society* 22, 127–138.

Chopra, PN (1969). *Who's Who of Indian Martyrs.* Ministry of Education and Youth Services, Government of India.

Noehden, GH (1824). Account of the banyan-tree, or *Ficus indica*, as found in the Ancient Greek and Roman authors. *Transactions of the Royal Asiatic Society of Great Britain and Ireland* 1, 119–132.

Sayeed, VA (2012). Arboreal wonder. *Frontline Magazine*, June 2012.

Strabo, *Geography*, trans. HL Jones (1932). Vol 7. Loeb Classical Library.

Thanos CA (1994). Aristotle and Theophrastus on plant–animal interactions, in M Arianoutsou & RH Groves (eds.), *Plant–Animal Interactions in Mediterranean-type Ecosystems*, 3–11. Kluwer Academic Publishers.

Thanos CA (2005). The geography of Theophrastus' life and of his botanical writings (Περι Φυτων). In: AJ Karamanos & CA Thanos (eds.), *Biodiversity and Natural Heritage in the Aegean*, Proceedings of the Conference 'Theophrastus 2000', Eressos-Sigri, Lesbos. (6–8 July 2000), Frangoudis, 113–131.

Theophrastus, *Enquiry into Plants*, trans. A Hort (1916). W. Heinemann.

Theophrastus, *De Causis Plantarum*, trans. B Einarson & GKK Link (1990). Harvard University Press.

CHAPTER FIVE: Botanical Monkeys

Barlow, HS (1993). Botanical Monkeys by E. J. H. Corner (book review). *Journal of Southeast Asian Studies* 24, 182–184.

Berg, CC & Corner, EJH (2005). *Ficus*—Moraceae. *Flora Malesiana*, Series I, 17, 1–730.

Burkhill, HM (1977). Introduction. *Gardens' Bulletin Singapore* 29, 1–2.

Colonial Office: Straits Settlements. (1945). Miscellaneous reports— British Military Administration, Malaya. British National Archives Reference No: CO 273/675/6.

Corner, EJH (1940). *Wayside Trees of Malaya*. Government Printing Office, Singapore.

Corner, EJH (1956). Merah the berok: A little hand among the trees. *The Straits Times*, 1 January 1956, 13.

Corner, EJH (1960). Taxonomic notes on *Ficus* L., Asia and Australasia. Sections 1-4 *Gardens' Bulletin Singapore* 17, 368–485.

Corner, EJH (1964). Royal Society Expedition to North Borneo 1961: Special Reports: 3. *Ficus* on Mt. Kinabalu. *Proceedings of the Linnean Society of London* 175, 37–39.

Corner, EJH (1965). Check-list of Ficus in Asia and Australasia with keys to identification. *The Gardens' Bulletin Singapore* 21, 1–186.

Sources

Corner, EJH (1967). *Ficus* in the Solomon Islands. *Philosophical Transactions of the Royal Society of London, Series B: Biological Sciences* 253, 23–159.

Corner, EJH (1969). *Ficus* (A discussion on the results of the Royal Society expedition to the British Solomon Islands Protectorate, 1965). *Philosophical Transactions of the Royal Society of London: Series B, Biological Science* 255, 567–570.

Corner, EJH (1981). *The Marquis, a Tale of Syonan-to*. Heinemann Asia.

Corner, EJH (1985). Essays on *Ficus*. *Allertonia* 4, 125–168.

Corner, EJH (1985). *Ficus* (Moraceae) and *Hymenoptera* (Chalcidoidea): Figs and their pollinators. *Biological Journal of the Linnean Society* 25, 187–195.

Corner, EJH (1992). *Botanical monkeys*. Pentland Press.

Corner, EJH (1993). I am part of all that I have met. In: S Isaac, JC Frankland, R Watling & AJ Whalley (eds.), *Aspects of Tropical Mycology*. Cambridge University Press, 1–14.

Corner, JK (2013). *My Father in his Suitcase: In Search of E. J. H. Corner, the Relentless Botanist*. Landmark Books.

Gudger, EW (1923). Monkeys trained as harvesters. Instances of a practice extending from remote times to the present. *Natural History Magazine*, May–June 1923.

Krishna, N & Amirthalingam, M (2014). *Sacred Plants of India*. Penguin.

Linnaeus, C (1753). *Species Plantarum*. Laurentius Salvius.

Mabberley, DJ (2000). A tropical botanist finally vindicated. *Gardens' Bulletin Singapore* 52, 1–4.

Mabberley, DJ & Lan, CK (eds.) (1977). Tropical Botany: Essays presented to E. J. H. Corner for his seventieth birthday, 1976. *Gardens' Bulletin Singapore* 29, 1–266.

Seemann, BC (1868). *Flora Vitiensis. A Description of the Plants of the Viti or Fiji Islands, with an Account of Their History, Uses, and Properties. Part 7*. L. Reeve.

Shanahan, M et al. (2001). Fig-eating by vertebrate frugivores: a global review. *Biological Reviews* 76, 529–572.

Vines, G (2002). King of the canopy. *New Scientist*, 21 December 2002.

Weiblen, G D & Clement, WL (2007). *Flora Malesiana*. Series I. Vol 17, Parts 1 & 2. *Edinburgh Journal of Botany* 64, 431–437.

Whitmore, T (1996). Obituary: Professor E. J. H. Corner. *The Independent*, 21 September 1996.

CHAPTER SIX: Sex & Violence in the Hanging Gardens

Ahmed, S et al. (2009). Wind-borne insects mediate directional pollen transfer between desert fig trees 160 kilometers apart. *Proceedings of the National Academy of Sciences* 106, 20342–20347.

Bain, A, Harrison, RD & Schatz, B (2014). How to be an ant on figs. *Acta Oecologica* 57, 97–108.

Berg, CC & Wiebes, JT (1992). *African Fig Trees and Fig Wasps*. Koninklijke Nederlandse Akademie van Wetenschappen.

Bleher, B et al. (2003). The importance of figs for frugivores in a South African coastal forest. *Journal of Tropical Ecology* 19, 375–386.

Burrows, J & Burrows, S (2003). *Figs of Southern & South-Central Africa*. Umdaus Press.

Compton, SG & Robertson, HG (1988). Complex interactions between mutualisms: ants tending Homopterans protect fig seeds and pollinators. *Ecology* 69, 1302–1305.

Cook, JM & Rasplus, J-Y (2003). Mutualists with attitude: coevolving fig-wasps and figs. *Trends in Ecology and Evolution* 18, 241–8.

Cook, JM et al. (2015). Fighting in fig-wasps: do males avoid killing brothers or do they never meet them? *Ecological Entomology* 40, 741–747.

Sources

Datwyler, SL & Weiblen, GD (2004). On the origin of the fig: phylogenetic relationships of Moraceae from ndhF sequences. *American Journal of Botany* 91, 767-777.

Deeble, M & Stone V (2005). *The Queen of Trees* [documentary film]. Flat Dog Productions Limited.

Galil, J & Eisikowitch, D (1968). On the pollination ecology of *Ficus sycomorus* in East Africa. *Ecology* 49, 259–269.

Galil, J & Eisikowitch, D (1974). Further studies on pollination ecology in *Ficus sycomorus* II. Pocket filling and emptying by *Ceratosolen arabicus* Mayr. *New Phytologist* 73, 515–528.

Ghara, M, Kundanati, L & Borges, R (2011). Nature's Swiss army knives: ovipositor structure mirrors ecology in a multitrophic fig-wasp community. *PLOS One* 6, e23642.

Herre, EA (1989). Coevolution of reproductive characteristics in 12 species of New World figs and their pollinator wasps. *Experientia* 45, 637–647.

Herre, EA, Jander, KC & Machado, CA (2008). Evolutionary ecology of figs and their associates: recent progress and outstanding puzzles. *Annual Review of Ecology, Evolution, and Systematics* 39, 439–458.

Jauharlina, J et al. (2012). Fig-wasps as vectors of mites and nematodes. *African Entomology* 20, 101–110.

Kerdelhué, C & Rasplus, J-Y (1996). Non-pollinating Afrotropical fig-wasps affect the fig-pollinator mutualism in *Ficus* within the subgenus *Sycomorus*. *Oikos* 75, 3–14.

Kinnaird, MF, O'Brien, TG & Suryadi, S (1999). The importance of figs to Sulawesi's imperiled wildlife. *Tropical Biodiversity* 6, 5–18.

Kissling, WD, Rahbek, C & Böhning-Gaese, K (2007). Food plant diversity as broad-scale determinant of avian frugivore richness. *Proceedings of the Royal Society, Series B: Biological Sciences* 274, 799–808.

Kjellberg, F et al. (2005). Biology, ecology and evolution of fig-pollinating wasps (*Chalcidoidea, Agaonidae*), in A Raman,

W Schaefer & Withers (eds.) *Biology, Ecology and Evolution of Gall-Inducing Arthropods*. Science Publishers, Inc., 539–572.

Korine, C, Kalko, EKV & Herre, EA (2000). Fruit characteristics and factors affecting fruit removal in a Panamanian community of strangler figs. *Oecologia* 123, 560–568.

Lambert, FR & Marshall, AG (1991). Keystone characteristics of bird-dispersed *Ficus* in a Malaysian lowland rain forest. *Journal of Ecology* 79, 793–809.

Leighton, M & Leighton, DR (1983). Vertebrate responses to fruiting seasonality within a Bornean rain forest. In: SL Sutton, TC Whitmore and AC Chadwick (eds.), *Tropical Rain Forest: Ecology and Management*, 181–196. Blackwell.

Machado, CA et al. (2001). Phylogenetic relationships, historical biogeography and character evolution of fig-pollinating wasps. *Proceedings of the Royal Society, Series B: Biological Sciences* 268, 685–694.

Machado, CA et al. (2005). Critical review of host specificity and its coevolutionary implications in the fig–fig-wasp mutualism. *Proceedings of the National Academy of Sciences* 102, 6558–6565.

Nason, JD, Herre, EA & Hamrick, JL (1998). The breeding structure of a tropical keystone resource. *Nature* 391, 685–687.

Rønsted, N et al. (2005). 60 million years of co-divergence in the fig–wasp symbiosis. *Proceedings of the Royal Society, Series B: Biological Sciences* 272, 2593–2599.

Rønsted, N et al. (2005). Reconstructing the phylogeny of figs (*Ficus*, Moraceae) to reveal the history of the fig pollination mutualism. *Symbiosis* 45, 45–56.

Shanahan, M et al. (2001). Fig-eating by vertebrate frugivores: a global review. *Biological Reviews* 76, 529–572.

Somjee, S (2014). The Ubuntu stratagem. Utu and peace sustaining heritage of Africa south of the Sahara. *African Peace Journal*.

Sources

Suleman, N, Raja, S & Compton, SG (2012). Only pollinator fig wasps have males that collaborate to release their females from figs of an Asian fig tree. *Biology Letters* 8, 344–346.

Terborgh, J (1986). Keystone plant resources in the tropical forest, in ME Soule (ed.) *Conservation Biology, the Science of Scarcity and Diversity*. Sinauer, 330–344.

Weiblen, GD (2002). How to be a fig-wasp. *Annual Review of Entomology* 47, 299–330.

Weiblen, GD (2004). Correlated evolution in fig pollination. *Systematic Biology* 53, 128–139.

Wiebes, JT (1976). A short history of fig-wasp research. *Gardens' Bulletin Singapore* 29, 207–232.

Wiebes, JT (1979). Co-evolution of figs and their insect pollinators. *Annual Review of Ecology and Systematics* 10, 1–12.

Willdenow, CL (1806). Observations on the genus *Ficus*, with the description of some new species. In: C Konig & J Sims (eds.), *Annals of Botany*, Volume II, 312–325. R. Taylor and Co.

CHAPTER SEVEN: Struggles for Existence

Beeler, C (2015). In quest for fidelity, a model from the animal kingdom. *Newsworks: The Pulse*, 12 February 2015.

Corner, EJH (1940). *Wayside Trees of Malaya*. Government Printing Office, Singapore.

Hadiprakarsa, Y-Y & Kinnaird, MF (2004). Foraging characteristics of an assemblage of four Sumatran hornbill species. *Bird Conservation International* 14, S53–S62.

Harrison RD et al. (2012). Evolution of fruit traits in *Ficus* subgenus *sycomorus* (Moraceae): To what extent do frugivores determine seed dispersal mode? *PLOS ONE* 7, e38432.

Harrison, RD et al. (2003). The diversity of hemi-epiphytic figs (*Ficus*; Moraceae) in a Bornean lowland rainforest. *Biological Journal of the Linnean Society* 78, 439–455.

Johns, AD (1987). The use of primary and selectively logged forest by Malaysian hornbills (Bucerotidae) and implications for their conservation. *Biological Conservation* 40, 179–190.

Kemp, AC & Woodcock, M (1995). *The Hornbills: Bucerotiformes*. Oxford University Press.

Kinnaird, MF & O'Brien, TG (2007). *The Ecology and Conservation of Asian Hornbills: Farmers of the Forest*. University of Chicago Press.

Kinnaird, MF, O'Brien, TG & Suryadi, S (1996). Population fluctuation in Sulawesi Red-knobbed Hornbill *Aceros cassidix*: tracking figs in space and time. *The Auk* 113, 431–440.

Laman, T (2009). High on hornbills. *National Wildlife*, February/March 2009.

Laman, T (2010). Hooked on hornbills. *Living Bird*, Autumn 2010.

Laman, T (2013). Borneo—My rain forest roots [online article], 17 January 2013. www.timlaman.com.

Laman, T (2014). Postcards from Borneo: A family adventure begins anew. Proof: Picture stories [online article], 1 July 2014. www.nationalgeographic.com.

Laman, TG & Weiblen, GD (1998). Figs of Gunung Palung National Park (West Kalimantan, Indonesia). *Tropical Biodiversity* 5, 245–97.

Laman, TG (1995). *Ficus stupenda* germination and seedling establishment in a Bornean rain forest canopy. *Ecology* 76: 2617–2626.

Laman, TG (1995). Safety recommendations for climbing rain forest trees with 'single rope technique'. *Biotropica* 27: 406–409.

Laman, TG (1995). The ecology of strangler fig seedling establishment. *Selbyana* 16, 223–229.

Sources

Laman, TG (1996). *Ficus* seed shadows in a Bornean rain forest. *Oecologia* 107, 347–355.

Laman, TG (1996). Specialization for canopy position by hemiepiphytic *Ficus* species in a Bornean rain forest. *Journal of Tropical Ecology* 12, 789–803.

Laman, TG (1996). The impact of seed harvesting ants (*Pheidole* sp. nov.) on *Ficus* establishment in the canopy. *Biotropica* 28, 777–781.

Laman, TG (1997). Borneo's strangler fig trees. *National Geographic* 191, 38–55.

Lambert, FR (1989). Fig eating by birds in a Malaysian lowland forest. *Journal of Tropical Ecology* 5, 401–412.

Lambert, FR (1989). Pigeons as seed predators and dispersers of figs in a Malaysian lowland forest. *Ibis* 131, 521–527.

Lee, HS et al. (2002). Floristic and structural diversity of mixed dipterocarp forest in Lambir Hills National Park, Sarawak, Malaysia. *Journal of Tropical Forest Science* 14, 379–400.

Lee, HS et al. (2002). The 52-hectare Forest Research Plot at Lambir Hills, Sarawak, Malaysia: Tree distribution maps, diameter tables and species documentation. Forest Department Sarawak & the Arnold Arboretum–CTFS Asia Program.

Leighton, M (1982). Fruit resources and patterns of feeding, spacing and grouping among sympatric Bornean hornbills (Bucerotidae). PhD thesis, University of California, Davis.

National Geographic photographer profile: Tim Laman [online article]. www.nationalgeographic.com.

Peart, DR (2003). The Road to Cabang Panti. Unpublished manuscript. Dartmouth College.

Poonswad, P & Tsuji, A (1994). Ranges of males of the Great Hornbill *Buceros bicornis*, Brown Hornbill *Ptilolaemus tickelli* and Wreathed Hornbill *Rhyticeros undulatus* in Khao Yai National Park, Thailand. *Ibis* 136, 79–86.

Porter Brown, N (2013). Paradise found. *Harvard Magazine* (January–February 2013), 62–65.

Shanahan, M & Compton, S (2001). Vertical stratification of figs and fig-eaters in a Bornean lowland rain forest: how is the canopy different? *Plant Ecology* 153, 121–132.

Shanahan, M & Compton. S (2000). Fig-eating by Bornean treeshrews: evidence for a role as seed dispersers. *Biotropica* 32, 759–764.

Shanahan, M (2000). *Ficus* seed dispersal guilds: ecology, evolution and conservation implications. PhD Thesis. University of Leeds.

Shanahan, M et al. (2001). Fig-eating by vertebrate frugivores: a global review. *Biological Reviews* 76, 529–572.

Wallace, AR (1863). The Bucerotidæ, or hornbills. *The Intellectual Observer*, June 1863, 309–316.

CHAPTER EIGHT: Goodbye to the Gardeners, Hello to the Heat

Ahmed, S et al. (2009). Wind-borne insects mediate directional pollen transfer between desert fig trees 160 kilometers apart. *Proceedings of the National Academy of Sciences* 106, 20342–20347.

Al-Khalaf, A, Quinnell, RJ, & Compton, SG (2015). Influence of temperature on the reproductive success of a fig-wasp and its host plant. *African Journal of Agricultural Research* 10, 1625–1630.

Bryan, JE et al. (2013). Extreme differences in forest degradation in Borneo: Comparing practices in Sarawak, Sabah, and Brunei. *PLOS ONE* 8, e69679.

Davenport, WH (2000). Hornbill carvings of the Iban of Sarawak, Malaysia. *RES: Anthropology and Aesthetics* 37, 127–146.

Fredriksson, GM, Danielsen, LS & Swenson, JE (2007). Impacts of El Niño related drought and forest fires on sun bear fruit resources in lowland dipterocarp forest of East Borneo. *Biodiversity and Conservation* 16, 1823–1838.

Sources

Gaveau, DLA et al. (2014). Four decades of forest persistence, clearance and logging on Borneo. *PLOS One* 9, e101654.

Harrison, RD & Shanahan, M (2005). Seventy-seven ways to be a fig: An overview of a diverse assemblage of figs in Borneo. In: DW Roubik, S Sakai, & AA Hamid (eds.), *Pollination, Ecology and the Rain Forest Canopy: Sarawak Studies*. Springer Verlag, 111–127.

Harrison, RD (2000). Repercussions of El Niño: Drought causes extinction and the breakdown of mutualism in Borneo. *Proceedings of the Royal Society, Series B: Biological Sciences* 267, 911–915.

Harrison, RD (2001). Drought and the consequences of El Niño in Borneo: a case study of figs. *Population Ecology* 43, 63–75.

Harrison, RD (2005). Figs and the diversity of tropical forests. *BioScience* 55, 1053–1064.

Harrison, RD (2005). A severe drought in Lambir Hills National Park. In: DW Roubik, S Sakai, & AA Hamid (eds.), *Pollination, Ecology and the Rain Forest Canopy: Sarawak Studies*. Springer Verlag, 51–64.

Harrison, RD (2011). Emptying the Forest: Hunting and the extirpation of wildlife from tropical nature reserves. *BioScience* 61, 919–924.

Harrison, RD et al. (2003). The diversity of hemiepiphytic figs (*Ficus*; Moraceae) in a Bornean lowland rainforest. *Biological Journal of the Linnean Society* 78, 439–455.

Janzen, DH (1979). How to be a fig. *Annual Review of Ecology and Systematics* 10, 13–51.

Jevanandam, N, Goh, AGR & Corlett, R (2013). Climate warming and the potential extinction of fig-wasps, the obligate pollinators of figs. *Biology Letters* 9, 20130041.

Johns, AD (1987). The use of primary and selectively logged forest by Malaysian hornbills (Bucerotidae) and implications for their conservation. *Biological Conservation* 40, 179–190.

Kinnaird, MF & O'Brien, TG (2007). *The Ecology and Conservation of Asian Hornbills: Farmers of the Forest*. University of Chicago Press.

Mohd-Azlan, J & Engkamat, L (2006). Camera trapping and conservation in Lambir Hills National Park, Sarawak. *The Raffles Bulletin of Zoology* 54, 469–475.

Nakagawa, M et al. (2000). Impact of severe drought associated with the 1997–1998 El Niño in a tropical forest in Sarawak. *Journal of Tropical Ecology* 16, 355–367.

Redford, K (1992). The Empty Forest. *BioScience* 42, 412–422.

Shanahan, M & Compton, S (2001). Vertical stratification of figs and fig-eaters in a Bornean lowland rain forest: how is the canopy different? *Plant Ecology* 153, 121–132.

Shanahan, M & Compton. S (2000). Fig-eating by Bornean treeshrews: evidence for a role as seed dispersers. *Biotropica* 32, 759–764.

Shanahan, M & Debski, I (2002). Vertebrates of Lambir Hills National Park, Sarawak, Malaysia. *Malayan Nature Journal* 56, 103–118.

Shanahan, M, 2000. *Ficus* seed dispersal guilds: ecology, evolution and conservation implications. PhD. Thesis. University of Leeds.

Shanahan, M et al. (2001). Fig-eating by vertebrate frugivores: a global review. *Biological Reviews* 76, 529–572.

CHAPTER NINE: From Dependence to Domination

AAAS (2009). Interview of Tim White by Ed Lempinen. SciPak / American Association for the Advancement of Science, 30 September 2009.

AAAS (2009). Oldest hominid skeleton unveiled. Press backgrounder 'Ardipithecus'. SciPak / American Association for the Advancement of Science, 1 October 2009.

Sources

Alba, DM et al. (2015). Miocene small-bodied ape from Eurasia sheds light on hominoid evolution. *Science* 350, aab2625.

Bassie-Sweet, K (2008). *Maya Sacred Geography and the Creator Deities*. University of Oklahoma Press.

Black, J (1998). *Reading Sumerian Poetry*. Cornell University Press.

Black, JA et al. (1998–2006). The Electronic Text Corpus of Sumerian Literature [online reference]. Faculty of Oriental Studies, University of Oxford. www.etcsl.orinst.ox.ac.uk.

Cappers, RTJ & Hamdy, R (2007). Ancient Egyptian plant remains in the Agricultural Museum (Dokki, Cairo). In RTJ Cappers (ed.) *Fields of Change: Progress in African Archaeobotany*. Barkhuis, 165–214.

Chan, H (2007). Survival in the Rainforest: Change and resilience among the Penan Vuhang of Eastern Sarawak, Malaysia. *Research Series in Anthropology*. University of Helsinki.

Denham, T (2007). Early fig domestication, or gathering of wild parthenocarpic figs? *Antiquity* 81, 457–461.

Dominy NJ et al. (2016). How chimpanzees integrate sensory information to select figs. *Interface Focus* 6, 20160001.

Galil, J & Eisikowitch, D (1968). On the pollination ecology of *Ficus sycomorus* in East Africa. *Ecology* 49, 259–269.

Galil, J (1968). An ancient technique for ripening sycamore fruit in East-Mediterranean countries. *Economic Botany* 22, 178–190.

Galil, J, Stein, M & Horovitz, A (1977). On the origin of the Sycamore Fig (*Ficus sycomorus* L.) in the Middle East. *Gardens' Bulletin Singapore* 29, 191–205.

Gopukumar, ST & Praseetha, PK (2015). *Ficus benghalensis* Linn— the sacred Indian medicinal tree with potent pharmacological remedies. *International Journal of Pharmaceutical Sciences Review and Research* 32, 223–227.

Handcock, PSP (1912). *Mesopotamian Archaeology: An introduction to the archaeology of Babylonia and Assyria*. Macmillan and Co.

Hopkin, M (2005). Ethiopia is top choice for cradle of *Homo sapiens* [online article]. *Nature*, 16 February 2005.

Ipulet, P (2007). Uses of genus *Ficus* (Moraceae) in Buganda region, Uganda. *African Journal of Ecology* 45, 44–47.

Janečka, JE et al. (2007). Molecular and genomic data identify the closest living relative of primates. *Science* 318, 792–794.

Janmaat, K et al. (2014). Wild chimpanzees plan their breakfast time, type, and location. *Proceedings of the National Academy of Sciences* 111, 16343–16348.

Janzen, JM (1978). *The Quest for Therapy in Lower Zaire*. University of California Press.

Janzen, JM (2012). Teaching the Kongo transatlantic. Newsletter, The African Diaspora Archaeology Network, Spring 2012.

Jolly-Saad, M-C et al. (2010). *Ficoxylon* sp., a fossil wood of 4.4 Ma (Middle Awash, Ethiopia). *Comptes Rendus Palevol* 9, 1–4.

Kislev, M, Hartmann, A & Bar-Yosef, O (2006). Early domesticated fig in the Jordan valley. *Science* 312, 1372–74.

Kislev, ME, Hartmann, A & Bar-Yosef, O (2006). Response to Comment on 'Early Domesticated Fig in the Jordan Valley'. *Science* 314, 1683.

Kunwar, RM & Bussman, RW (2006). *Ficus* (fig) species in Nepal: a review of diversity and indigenous uses. *Lyonia* 11, 85–97.

Lev-Yadun, S et al. (2006). Comment on 'Early domesticated fig in the Jordan Valley'. *Science* 314, 1683.

Liu, W et al. (2015). The earliest unequivocally modern humans in southern China. *Nature* 526, 696–699.

Lovejoy, CO (2009). Re-examining human origins in light of *Ardipithecus ramidus*. *Science* 326, 74–74e1–74e8.

Martin, EA et al. (2009). Conservation value for birds of traditionally managed isolated trees in an agricultural landscape of Madagascar. *Biodiversity and Conservation* 18, 2719–2742.

Sources

Masi, S et al. (2012). Unusual feeding behavior in wild great apes, a
window to understand origins of self-medication in humans: Role
of sociality and physiology on learning process. *Physiology &
Behavior* 105, 337–349.

Maspero, G (1903). *History of Egypt, Chaldea, Syria, Babylonia and
Assyria*. The Grolier Society.

Maundu, P et al. (2001). Ethnobotany of the Loita Maasai: Towards
Community Management of the Forest of the Lost Child.
Experiences from the Loita Ethnobotany Project. People and Plants
Working Paper Number 8. UNESCO, Paris.

McDougall, I, Brown, FH & Fleagle, JG (2005). Stratigraphic
placement and age of modern humans from Kibish, Ethiopia.
Nature 433, 733–736.

Merlin, M (2015). *People and Plants of Micronesia: Database of
economic plants of Micronesia*. University of Hawai'i at Mānoa.

Morales, J & Delgado, T (2007). Figs and their importance in the
prehistoric diet in Gran Canaria Island (Canary Isles). In: RTJ
Cappers (ed.) *Fields of Change: Progress in African
Archaeobotany*. Barkhuis, 77–85.

Nicholson, PT & Shaw, I (eds.) (2000). *Ancient Egyptian Materials
and Technology*. Cambridge University Press.

Normand, E & Boesch, C (2009). Sophisticated Euclidean maps in
forest chimpanzees. *Animal Behaviour* 77, 1195–1201.

Normand, E et al. (2009). Forest chimpanzees (*Pan troglodytes verus*)
remember the location of numerous fruit trees. *Animal Cognition*
12, 797–807.

Oakley, KP (1932). Woods used by the ancient Egyptians. *Analyst* 57,
158–159.

O'Leary, MA et al. (2013). The placental mammal ancestor and the
post-K–Pg radiation of placentals. *Science* 339, 662–667.

Plutarch. *Moralia*, trans FC Babbitt (1931). Harvard University Press.

Porteous, A.(2005). *The Lore of the Forest*. Cosimo Classics.

Prentice, R (2010). *The Exchange of Goods and Services in Pre-Sargonic Lagash*. Alter Orient und Altes Testament 368. Ugarit-Verlag, Münster.

Renne, P et al. (2015). State shift in Deccan volcanism at the Cretaceous-Paleogene boundary, possibly induced by impact. *Science* 350, 76–78.

Sayers, K, Raghanti, MA & Lovejoy, CO (2012). Human evolution and the chimpanzee referential doctrine. *Annual Review of Anthropology* 41, 119–138.

Schulte, P et al. (2010). The Chicxulub asteroid impact and mass extinction at the Cretaceous–Paleogene boundary. *Science* 327, 1214–1218.

Shi, J et al. (2014). An ethnobotanical study of the less known wild edible figs (genus *Ficus*) native to Xishuangbanna, Southwest China. *Journal of Ethnobiology and Ethnomedicine* 10, 68.

Stanford, CB (2012). Chimpanzees and the behavior of *Ardipithecus ramidus*. *Annual Review of Anthropology* 41, 139–149.

Theophrastus. *Enquiry into Plants*, trans. A. Hort (1916). W. Heinemann.

Watts, DP et al. (2012). Diet of chimpanzees (*Pan troglodytes schweinfurthii*) at Ngogo, Kibale National Park, Uganda, 1. diet composition and diversity. *American Journal of Primatology* 74, 114–129.

White, TD et al. (2009). *Ardipithecus ramidus* and the paleobiology of early hominids. *Science* 326, 75–86.

White, TD et al. (2015). Neither chimpanzee nor human, *Ardipithecus* reveals the surprising ancestry of both. *Proceedings of the National Academy of Sciences* 112, 4877–4884.

White, TD, Suwa, G & Asfaw, B (1994). *Australopithecus ramidus*, a new species of early hominid from Aramis, Ethiopia. *Nature* 371, 306–312.

Sources

WoldeGabriel, G et al. (2009). The geological, isotopic, botanical, invertebrate, and lower vertebrate surroundings of *Ardipithecus ramidus*. *Science* 326, 65–65e1–65e5.

Wrangham, RW et al. (1993). The value of figs to chimpanzees. *International Journal of Primatology* 14, 243–256.

Wrangham, RW et al. (1994). Seed dispersal by forest chimpanzees in Uganda. *Journal of Tropical Ecology* 10, 355–368.

Zohary, D, Hopf, M & Weiss, E (2012). *Domestication of Plants in the Old World*. (4th edn). Oxford University Press.

CHAPTER TEN: The War of the Trees

American Public Media (2011). Wangari Maathai—Planting the Future [transcript]. *On Being with Krista Tippett*. 29 September 2011.

Anderson, D (2005). *Histories of the Hanged: The Dirty War in Kenya and the End of Empire*. Weidenfeld.

Beech, MWH (1913). A ceremony at a mugumu or sacred fig-tree of the A-Kikuyu of East Africa. *Man* 13, 86–89.

Beech, MWH (1913). The sacred fig-tree of the A-Kikuyu of East Africa. *Man* 13, 4–6.

Berman, B & Lonsdale, J (1992). *Unhappy Valley: Conflict in Kenya & Africa. Book 2: Violence and Ethnicity*. Ohio University Press.

Biles, P (2012). Mau Mau massacre documents revealed [online article]. BBC Online (30 November 2012).

Cagnolo, C (1933). *The Akikuyu. Their Customs, Traditions and Folklore*. Mathari Press.

Chappell, S (2011). Airpower in the Mau Mau conflict: the government's chief weapon. *RUSI Journal* 156, 64–70.

Davidson, B (1994). The motives of the Mau Mau. *London Review of Books*, (24 February 1994), 12.

Films Media Group (2009). *Wangari Maathai: For Our Land* [documentary film].

Gathogo, J (2013). Environmental management and African indigenous resources: echoes from Mutira Mission, Kenya (1912–2012). *Studia Historiae Ecclesiasticae* 39, 33–56.

Henderson, I & Goodhart, P (1958). *The Hunt for Kimathi*. Hamish Hamilton.

Hewitt, P (2008). *Kenya Cowboy: A Police Officer's Account of the Mau Mau Emergency*. 30 Degrees South Publishers.

Hughes, L (2002). Moving the Maasai: A colonial misadventure. DPhil thesis. University of Oxford.

Huxley, E (1991). *Nine Faces of Kenya*. The Harvill Press.

Kamenju wa Mwangi, JW (2008). Gikuyu origins [online article]. 13 November 2008). www.mukuyu.wordpress.com.

Karangi, MM (2008). Revisiting the roots of Gĩkũyũ culture through the sacred Mũgumo tree. *Journal of African Cultural Studies* 20, 117–132.

Karanja, J (2009). *The Missionary Movement in Colonial Kenya: The Foundation of Africa Inland Church*. Cuvillier Verlag.

Kenyatta, J (1965). *Facing Mt. Kenya: The Tribal Life of the Gikuyu*. Vintage.

Leakey, LSB (1977). *The Southern Kikuyu before 1903. Volume 1*. Academic Press.

Maathai, W (2006). *Unbowed: A Memoir*. Alfred A. Knopf.

Maathai, W (2010). *Replenishing the Earth: Spiritual Values for Healing Ourselves and the World*. Doubleday.

Merton, L & Dater, A (2008). *Taking Root: The Vision of Wangari Maathai* [documentary film]. Marlboro Productions.

Muiruri, S (1999). MPs, Maathai beaten at forest. *Daily Nation*, 9 January 1999.

Mwangi, E (1998). Colonialism, self-governance and forestry in Kenya: Policy, practice and outcomes. *Research in Public Affairs*. University of Indiana.

Sources

National Assembly (Kenya) (1995). Parliamentary Debate. Points of Order: Protection of mugumo tree in Thika. *Kenya National Assembly Official Record (Hansard)*, 20 July 1995, 1631–1632.

National Assembly (Kenya) (1996). Parliamentary Debate. Question 022: Maintenance of worshipping places. *Kenya National Assembly Official Record (Hansard)*, 9 May 1996, 737–738.

National Assembly (Kenya) (2002). Parliamentary Debate. *Kenya National Assembly Official Record (Hansard)*, 6 August 2002, 2082.

Ng'ang'a, W (2006). *Kenya's Ethnic Communities: Foundation of the Nation*. Gatundu Publishers Limited.

Ngugi, T (2013). Mugumo trees and the odd ambivalence of Kibaki. *The East African*, 19 January 2013.

Njagih, M (2010). KWS to market Kimathi tree 'mailbox' as historic tourist attraction [online article]. *Standard Digital*, 3 June 2010.

Njagih, M (2010). Secrets of old tree that was Mau Mau post office. *East African Standard*, 3 June 2010.

Nthamburi, Z (ed.) (1991). *From Mission to Church: A Handbook of Christianity in East Africa*. Uzima Press.

Ofcansky, TP & Maxon, RM (2000). *Historical Dictionary of Kenya*. Scarecrow Press.

Overton, JD (1990). Social control and social engineering: African reserves in Kenya 1895–1920. *Environment and Planning D: Society and Space* 8, 163–174.

Sandgren, DP (1982). Twentieth century religious and political divisions among the Kikuyu of Kenya. *African Studies Review* 25, 195–207.

Truth, Justice and Reconciliation Commission (2013). *The Final Report of the Truth Justice and Reconciliation Commission of Kenya*.

Walker, ES (1962). *Treetops Hotel*. Robert Hale Publishing.

Wamagatta, EN (2009). *The Presbyterian Church of East Africa: An Account of Its Gospel Missionary Society Origins, 1895–1946*. Peter Lang Publishing.

CHAPTER ELEVEN: The Testimony of Volcanoes

Ahmed, S et al. (2009). Wind-borne insects mediate directional pollen transfer between desert fig trees 160 kilometers apart. *Proceedings of the National Academy of Sciences* 106, 20342–20347.

Anon. (1885). The results of the Krakatoa eruption. *Science* 6: 291–293.

Ball, EE & Johnson, RW (1976). Volcanic history of Long Island, Papua New Guinea. In: RW Johnson (ed) *Volcanism in Australasia*. Elsevier.

Blong, RJ (1982). *The Time of Darkness: Local Legends and Volcanic Reality in Papua New Guinea*. University of Washington Press.

Compton, SG et al. (1988). The colonization of the Krakatau islands by fig wasps and other chalcids (Hymenoptera, Chalcidoidea). *Philosophical Transactions of the Royal Society B* 322, 459–490.

Cook, S, Singadan, R & Thornton, IWB (2001). Colonization of an island volcano, Long Island, Papua New Guinea, and an emergent island, Motmot, in its caldera lake. IV. Colonization by non-avian vertebrates. *Journal of Biogeography* 28, 1353–1363.

Dampier, W (1729). *A Continuation of a Voyage to New Holland, Etc. in the Year 1699 by William Dampier*. John and James Knapton.

Guevara, S, Laborde, J & Sanchez-Rio, G (2006). Rain forest regeneration beneath the canopy of fig trees isolated in pastures of Los Tuxtlas, Mexico. *Biotropica* 36, 99–108.

Hefferan, D & Hess, B (2013). Can *Ficus* sp. forests be restored through vegetative propagation? [research poster]. Drake University.

Sources

Johnson, RW (2013). *Fire Mountains of the Islands. A History of Volcanic Eruptions and Disaster Management in Papua New Guinea and the Solomon Islands.* The Australian National University.

Lomáscolo, SB et al. (2010). Dispersers shape fruit diversity in *Ficus* (Moraceae). *Proceedings of the National Academy of Sciences* 107, 14668–14672.

New, TR, Smithers, CN & Marshall, AT (2005). Ian Walter Boothroyd Thornton (1926-2002). *Historical Records of Australian Science* 16, 91–106.

Schipper et al. (2001). Colonization of an island volcano, Long Island, Papua New Guinea, and an emergent island, Motmot, in its caldera lake. III. Colonization by birds. *Journal of Biogeography* 28, 1339–1352.

Shanahan, M et al. (2001). Colonization of an island volcano, Long Island, Papua New Guinea, and an emergent island, Motmot, in its caldera lake. V. Colonization by figs (*Ficus* spp.), their dispersers and pollinators. *Journal of Biogeography* 28, 1365–1377.

Shilton, LA & Whittaker, RJ (2010). The role of pteropodid bats in re-establishing tropical forests on Krakatau. In: *Island Bats: Evolution, Ecology, and Conservation.* University of Chicago Press, pp. 176–215.

Shilton, LA et al. (1999). Old World fruit bats can be long-distance seed dispersers through extended retention of viable seeds in the gut. *Proceedings of the Royal Society of London*, B 266, 219–223.

Smithsonian Institution National Museum of Natural History Global Volcanism Program: Long Island [online report]. http://volcano.si .edu/volcano.cfm?vn=251050.

Thornton, IWB (1996). *Krakatau: The Destruction and Reassembly of an Island Ecosystem.* Cambridge, MA: Harvard University Press, 346.

Thornton, IWB, Compton, SG & Wilson, CN (1996). The role of animals in the colonization of the Krakatau islands by fig trees (*Ficus* species). *Journal of Biogeography* 23, 577–592.

Thornton, IWB (2001). Colonization of an island volcano, Long Island, Papua New Guinea, and an emergent island, Motmot, in its caldera lake. I. General introduction. *Journal of Biogeography* 28, 1299–1310.

Zahawi, RA (2008). Instant trees: Using giant vegetation stakes in tropical forest restoration. *Forest Ecology and Management* 255, 3013–3016.

CHAPTER TWELVE: Once Destroyed, Forever Lost?

Blakesley, D & Elliott, S (2003). Thailand, restoration of seasonally dry tropical forest using the Framework Species Method [online report]. Forest Restoration Research Unit, Chiang Mai University.

Elliott, S & Kuaraksa, C (2008). Producing framework tree species for restoring forest ecosystems in northern Thailand. *Small-Scale Forestry* 7, 403–415.

Elliott, S et al. (2003). Selecting framework tree species for restoring seasonally dry tropical forests in northern Thailand based on field performance. *Forest Ecology and Management* 184, 177–191.

Elliott, S, Anusarnsunthorn, V & Blakesley, D (1998). *Forests for the future: Growing and Planting Native Trees for Restoring Forest Ecosystems*. Within Design Co. Ltd, Chiang Mai.

Goosem, SP & Tucker, NIJ (1995). *Repairing the Rainforest: Theory and practice of rainforest re-establishment in North Queensland's wet tropics*. Wet Tropics Management Authority, Cairns, Australia.

Hansen, MC et al. (2013). High-resolution global maps of 21st-century forest cover change. *Science* 342, 850–853.

Kim, D-H, Sexton, JO, & Townshend, JR (2015). Accelerated deforestation in the humid tropics from the 1990s to the 2000s. *Geophysical Research Letters* 42, 3495–3501.

Kuaraksa, C & Elliott, S (2012). The use of Asian *Ficus* species for

restoring tropical forest ecosystems. *Restoration Ecology* 21,
86–95.

Pakkad, G et al. (2001). Forest restoration planting in northern
Thailand. *Proceedings of the SE-Asian Moving Workshop on
Conservation, Management and Utilisation of Forest Genetic
Resources*, Thailand, 25 February 2001.

Sinhaseni, K (2008). Natural establishment of tree seedlings in forest
restoration trials at Ban Mae Sa Mai. Chiang Mai Province. MSc
Thesis, Graduate School, Chiang Mai University.

Toktang, T (2005). The effects of forest restoration on the species
diversity and composition of a bird community in Doi Suthep-Pui
National Park, Thailand, from 2002 to 2003. MSc Thesis,
Department of Biology, Faculty of Science, Chiang Mai University.

Toktang, T, Elliott, S & Gale, G (2005). The effects of forest restoration on the species diversity and composition of a bird community
in Doi Suthep-Pui National Park. *Natural History Bulletin of the
Siam Society* 53, 156–157.

EPILOGUE: A Wedding Invitation

Anadolu Agency (2016). Muğla's historic structures threatened by
trees, plants. *Hurriyet Daily News* 19 February 2016.

Cook, BI et al. (2012). Pre-Columbian deforestation as an amplifier of
drought in Mesoamerica. *Geophysical Research Letters* 39, L16706.

Cooke, K (2016). Climate helped trigger Angkor's fall [online article].
climatenewsnetwork.net, 4 March 2016.

Cottee-Jones, HEW (2014). Isolated *Ficus* trees and conservation in
human-modified landscapes. DPhil. University of Oxford.

Cottee-Jones, HEW (2015). Isolated *Ficus* trees deliver dual
conservation and development benefits in a rural landscape. *Ambio*
44, 678–684.

Cottee-Jones, HEW et al. (2015). Are protected areas required to maintain functional diversity in human-modified landscapes? PLOS One 10, e0123952.

Cottee-Jones, HEW et al. (2016). The importance of *Ficus* (Moraceae) trees for tropical forest restoration. *Biotropica* doi: 10.1111/btp.12304.

Dixit, Y, Hodell, DA & Petrie, CA (2014). Abrupt weakening of the summer monsoon in northwest India 4100 years ago. *Geology* 42, 339–342.

Marris, E (2014). Two-hundred-year drought doomed Indus Valley Civilization [online article]. nature.com, 3 March 2014.

Masson, C (1842). *Narrative of Various Journeys in Balochistan, Afghanistan and the Panjab Including a Residence in Those Countries from 1826 to 1838*. Richard Bentley.

Lansky, EP (2008). *Ficus* spp. (fig): Ethnobotany and potential as anticancer and anti-inflammatory agents. *Journal of Ethnopharmacology* 119, 195–213.

Lansky, EP & Paavilainen, HM (2010). *Figs: The Genus* Ficus *(Traditional Herbal Medicines for Modern Times)*. CRC Press.

Turner, BL & Sabloff, JA (2012). Classic Period collapse of the Central Maya Lowlands: Insights about human–environment relationships for sustainability. *Proceedings of the National Academy of Sciences* 109, 13908–13914.

·About the Author

Mike Shanahan is a freelance writer with a doctorate in rainforest ecology. He has lived in a national park in Borneo, bred endangered penguins, investigated illegal bear farms, produced award-winning journalism and spent several weeks of his life at the annual United Nations climate change negotiations. He is interested in what people think about nature and our place in it. His freelance journalism includes work published by *The Economist*, *Nature*, *The Ecologist* and *Ensia*, and chapters of *Dry: Life without Water* (Harvard University Press, 2006), *Climate Change and the Media* (Peter Lang Publishing, 2009) and *Culture and Climate Change: Narratives* (Shed, 2014). He is the illustrator of *Extraordinary Animals* (Greenwood Publishing Group, 2007) and maintains a blog called *Under the Banyan*.

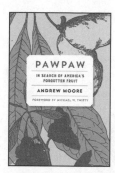
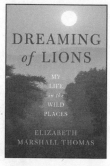

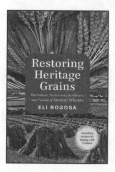